Computational Fluid Dynamics for Built and Natural Environments

Zhiqiang (John) Zhai

Computational Fluid Dynamics for Built and Natural Environments

 Springer

Zhiqiang (John) Zhai
University of Colorado at Boulder
Boulder, CO, USA

ISBN 978-981-32-9822-4   ISBN 978-981-32-9820-0   (eBook)
https://doi.org/10.1007/978-981-32-9820-0

This Springer imprint is published by the registered company Springer Nature Singapore Pte Ltd.
The registered company address is: 152 Beach Road, #21-01/04 Gateway East, Singapore 189721, Singapore

To My Beloved Father

Daozhong Zhai

*A Mechanical Engineer and Practitioner
(1944–2014)*

Contents

Chapter 1
Introduce CFD

1.1 What Is Fluid Mechanics

The essence of world and human life is fluids and flow. We all pass through life surrounded—and even sustained—by the flow of fluids. Blood moves through the vessels in our bodies, and air flows into our lungs. Our vehicles move through our planet's blanket of air or across its lakes and seas, powered by still other fluids, such as fuels and oxidizers, which mix in the combustion chambers of engines. The fluid flow indeed occurs in every aspect of our lives, such as:

- breathing, coughing, and sneezing;
- drinking, cooking, and digesting;
- delivering medicine in body;
- washing and drying clothes;
- swimming, biking, surfing, sailing, and parachuting;
- smoking, extinguishing a fire with water;
- heating, cooling or ventilating a room, etc.

Many of the environmental or energy-related issues we face today cannot possibly be confronted without detailed knowledge of the mechanics of fluids. Fluid mechanics is *the science of fluids (e.g., air, water, steam, oil, blood, etc.) and their behaviors at rest (statics) or in motion (dynamics) as well as the interaction of fluids with solids or other fluids.* Figure 1.1 illustrates such fluid statics and dynamics, which involves the flow mechanisms of lake water and the interactions between water and air, between water and the land bank, and between water and the boat. Another example is high-rise buildings or suspension bridges confronting a constant or instantaneous wind.

Fluid mechanics is a fundamental field that is the basis of many important industry and research topics such as aeronautics and astronautics, oil exploration, transportation tools (airplane, high-speed train, submarine, etc.), building ventilation, and human body metabolic regulation, etc. Well-known scientists have developed basic principles of fluid mechanics as outlined in Table 1.1. The development of

© Springer Nature Singapore Pte Ltd. 2020
Z. Zhai, *Computational Fluid Dynamics for Built
and Natural Environments,* https://doi.org/10.1007/978-981-32-9820-0_1

Fig. 1.1 Illustration of fluid statics and dynamics

Table 1.1 Historical milestones in fluid mechanics

Year	Scientist	Main contribution
250 B.C.	Archimedes	Buoyancy of floating bodies
1700	Isaac Newton	Linear law of viscosity
1738	Daniel Bernoulli	Energy law for incompressible fluids
1827	Claude Louis Marie Navier	Navier-Stokes equations for fluid
1845	George Gabriel Stokes	Motion and friction
1883	Osborn Reynolds	Reynolds number for pipe flows
1890	Lord Rayleigh	Dimensionless analysis
1904	Ludwig Prandtl	Boundary layer theory

fluid mechanics greatly fosters the developments of the other fields. However, the research of fluid flows is most challenging due to the sensitivity, randomness and uncertainty of flows, as well as the special properties of different fluids. It requires not only the strong physics background of researchers but also the skilled mathematics capability (Fig. 1.2).

1.2   How to Study Fluid Mechanics

Mechanics of fluid flow can be studied in three approaches:

- *Experimentally: experimental fluid mechanics.* This includes various full-scale and small-scale mock-up experiments, under either controlled or real operating

Frenchman, Claude Navier   Irishman, George Stokes
(1785-1836)   (1819-1903)

Fig. 1.2 Photos of Navier and Stokes

conditions of systems in study. Certain similarity principles need be met before the results obtained from a mock-up experiment can be generalized to real-world applications.

- *Analytically: theoretical fluid mechanics.* Analytical solutions may be obtained for some special flow problems. Most theoretical analyses of fluid mechanics can only be conducted for simple or simplified cases, such as steady and one-dimensional problems. Proper simplification or approximation is the key for a successful theoretical analysis.
- *Numerically: computational fluid mechanics.* This approach counts on computers (sometimes, supercomputers) to solve the complicated and highly non-linear governing equations of fluid flow using various numerical techniques. Computational fluid mechanics is typically called as *computational fluid dynamics (CFD)* as the approach is mostly powerful in handling dynamic behaviors in fluid flow.

The following paragraphs compare the advantages and disadvantages of these approaches, assuming they are performed by a researcher with adequate knowledge and skill in each area. Obviously, these methods complement each other, and an appropriate and integrated use of these research methods will ultimately lead to substantial developments in fluid mechanics and its relevant disciplines.

Advantages of the Experimental Approach

- **Physics**: Experiments may reveal the actual (and new) physics that is not discovered or understood with the current theoretical models. The new phenomena discovered will thus stimulate the new developments in fundamental theories.

- **Reality**: Experiments may provide first-hand information on flow characteristics under real conditions, which are usually the result of many comprehensive, influencing factors. Experimental results provide good resources to verify and validate various new theories and models developed upon a variety of approximations.

Disadvantages of the Experimental Approach

- **Cost**: Most physical experiments (either field or lab testing) can be very expensive in terms of facility construction, instrumentation, labor, and study time. Different types of instruments need to be acquired, allocated, calibrated, and monitored at the same location in a flow domain if multiple variables are to be measured.
- **Data**: Associated with the cost, only a limited amount of discrete data can be collected within a continuous space (such as a building). The scarceness of the data can provide challenges to make a complete understanding and analysis of the entire flow mechanisms.
- **Complication**: The comprehensive interactions among various influence factors in a physical experiment may sometimes prevent a clear understanding of the causes and consequences of the individual flow mechanisms involved.
- **Representation**: The testing results obtained in an experimental setup may only be used to reveal the physics observed in similar environments. Caution is required to generalize the experimental findings for other scenarios. Most laboratory experiments concentrate on exploring the fundamental mechanisms of flow, thus focusing on cases with simple geometries and flow conditions that can significantly differ from real applications.
- **Special conditions**: Due to the safety and health concerns, experiments with or under extreme conditions (e.g., with toxic or rotten materials, or under high pressure, extreme temperature conditions) will either require special handling of experimental facilities and instruments or use substitutive testing materials and/or conditions. This may thus result in significant cost increase or unrealistic testing environments.

Advantages of the Analytical Approach

- **Physics**: Analytical solutions, if acquired properly, can provide a straightforward and important insight on principal flow physics, which is critical for developing new flow assumptions, theories, and models. Analytical solutions often reveal the most elemental aspects of flow mechanisms with super-simple (simplified) problem setups.
- **Accuracy**: Due to the simple nature of problems studied, some analytical solutions may represent the exact solutions to the problems. These analytical solutions are of great value for validating new theories and models proposed.
- **Cost**: There is almost no monetary cost to conduct an analytical study for any flow problem, if the time cost is not considered that could be very significant depending on the capability of a researcher and the complexity of the problem.

Disadvantages of the Analytical Approach

- **Simplification**: Analytical approach can only be used for very few simple flow problems (such as, one-dimension, in-viscid flow etc.). Obtained analytical solutions may have very limited applicable ranges. Significant approximations are often required to establish an analytically solvable mathematic system, which can be largely different from real flow conditions.

Advantages of the Computational Approach

- **Relatively low cost**: Thanks to the rapid development in computer industry, the computational approach is less expensive in investment and can obtain more informative results with much shorter time. The costs are likely to decrease as computers become more powerful. For most studies, the cost of CFD simulation is almost negligible when compared to the experimental approach, whether on-site or mock-up experiments.
- **Speed**: The computational approach can be executed in a short period of time (ranging from a few seconds to a few days depending on the physics of the problem and the resolution requirement of the solution). Quick turnaround means engineering data can be introduced early in various decision-making processes.
- **Ability to simulate real conditions**: The computational approach provides the ability to theoretically simulate any physical condition, especially those that cannot be (easily) tested in experiments, e.g. hypersonic flow. CFD can effectively and safely model the situations under extreme or ideal conditions, such as, extreme-hot/cold and high-toxic scenarios, in which the measurement is usually very difficult or even impossible.
- **Ability to simulate ideal conditions**: The computational approach allows great control over the physical process, and provides the ability to isolate specific phenomena for study. For instance, a heat transfer process can be idealized with adiabatic, constant heat flux, or constant temperature boundaries. One can deliberately study the influence of a particular design feature on the whole system performance by adjusting this specific parameter while keeping others unchanged in the CFD simulation.
- **Comprehensive information**: The computational approach allows an analyst to examine a large number of locations in the domain of interest, and yields a comprehensive set of flow parameters (e.g., detailed distributions of air velocity, pressure, temperature, moisture, and contaminant concentrations etc.) for examination, mostly under a single computation. The information allows one to have a global knowledge of flows, rather than limited observations based on a few of measurement points.
- **Operation easiness**: Thanks to the attention and development in intelligent graphic user interface (GUI) technologies, a CFD user can easily change and test different modeling scenarios once the base model is well built and validated. In many

commercial CFD programs, the above operation may be as simple as clicking a button.

Disadvantages of the Computational Approach

- **Accuracy**: The accuracy of computational results is strongly dependent on whether the flow governing equations solved by a computer can correctly describe the flow physics (such as turbulence). In addition, developing an appropriate computer model of a reality sometimes is an art requiring profound knowledge, prior experience, and creativity of handling similar problems. Simplification processes of complex real objects into computer recognizable models will largely influence the accuracy of modeling results.

  Accurate prediction of airflow, temperature and contaminant concentration distributions requires deep understanding of physics of flows in the domain. For example, for building indoor environment modeling, since the airflow ranges from laminar to turbulent flow, a comprehensive airflow model considering both laminar and turbulence effect is desired. Although there are a number of turbulence models available nowadays, a universal model that is able to describe diversity of flow regimes in and around buildings is still not available. Meanwhile, the buoyancy and near-wall effect impose more challenges on the turbulence models. The distribution of air-phase contaminant concentration, although mainly determined by the airflow patterns, needs special models to handle the correlation between the fluctuation of concentration and airflow. Different contaminant sources may have different behaviors and need different models. If the contaminants are in liquid or solid particle phase, the problem is becoming the two-phase or multi-phase flow. Therefore, profound knowledge on various flow models for accurate CFD simulation is always desired.

  The accuracy of CFD prediction is highly sensitive to the boundary conditions supplied (assumed) by the user. These boundary conditions are crucial for the accuracy of the CFD results. The boundary conditions specified in CFD can be obtained from measurements. But most of them are based on empirical data or even experienced guess. The circumstance may become more challenging when time-varying boundary conditions are required for an unsteady calculation, in which the dynamic measurement data is usually unavailable and even the estimate is difficult to make.

- **Numerics**: Continuous space and time domains in physics must be discretized into discrete systems before a computer code can recognize and process. Various numerical schemes and methods are utilized during the process, which may lead to unstable, un-converged, and unrealistic results if handled improperly.

  Since the flow governing equations are highly non-linear and strongly self-coupled, CFD applies numerical methods, such as the finite volume method (FVM), to discretize the partial differential equations (PDE) and obtain the corresponding algebraic equations that can be solved iteratively. The numerical approximation and iterative calculation may introduce various uncertainties and instabilities. For instance, the high-order-accuracy differencing convection scheme may bring significant instability into the computation. That is why many state-of-the-art algorithms and techniques are created to ensure the calculations toward a convergent and stable direction, such as false-time-step and relaxation factor methods. However, when more sophisticated mathematical models (e.g. the Reynolds stress turbulence model) and numerical techniques (e.g., the multi-grid algorithm) are developed and used to handle the complex problems, the numerical stability and convergence problem is always of big concern.

1.3 What Is CFD?

With the rapid development of computer science and numerical techniques, numerical simulation of reality has been playing an increasingly important role. Computational Fluid Dynamics (CFD) is the field of numerical simulation of flow-related problems, typically using a computer.

With the assistance of computers and numerical algorithms, CFD is solving the flow governing equations (i.e., the fundamental conservation equations in mass, momentum, and energy) to predict what will happen, quantitatively, when fluids flow, often with the complications of:

- simultaneous flow of heat
- mass transfer (e.g., perspiration, dissolution)
- phase change (e.g., melting, freezing, boiling)
- chemical reaction (e.g., combustion, rusting)
- mechanical movement (e.g., of pistons, fans, rudders)
- stresses in and displacement of immersed or surrounding solids.

CFD results, after validations and verifications, can then be used to understand physics, improve designs, optimize systems, guide procedures, and influence decision-makings.

1.4 How Old Is CFD?

The early beginnings of CFD were in the 1960s, mostly moving along with the development pace of computer industry. Its first successes came to prominence in the 1970s, while the creation of the CFD-service industry started in the 1980s. In 1990s, the CFD industry expanded significantly due to the initial deployment of personal computers (mostly in research entities though). Expansion continued in the Second Millennium as commercial CFD packages developed easy GUIs (graphic user interface) and compatible connections with those for CAD and solid-stress analysis.

As a demonstration, Fig. 1.3 reveals the trend of using CFD in building indus-

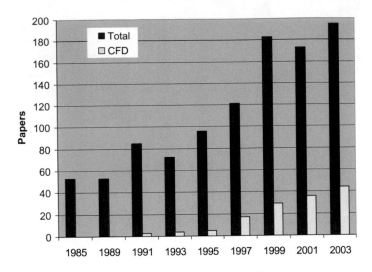

Fig. 1.3 Papers presented at earlier IBPSA conferences (Zhai 2006)

try between 1980s and 2000s by analyzing the numbers of CFD-relevant papers published in the proceedings of the biannual International Conference on Building Simulation of IBPSA (The International Building Performance Simulation Association), one of the most premier events in the field with a focus on computer simulation in building. It is noted that before 1997, CFD was new to most building designers and engineers and still at the stage of accumulating credentials by validating CFD with building experiments and discussing appropriate boundary conditions for building simulation. With the development of computer capacity and well establishment of CFD reputation, CFD had gained more and more attentions in 2000s due to its advantages mentioned above. It thus has been growingly used for various building projects. Buildings and systems modeled in CFD have become more and more sophisticated, while less knowledge of fluid mechanics and building science is required to conduct CFD simulations due to smart GUIs of commercial CFD programs. This, therefore, results in the necessity of developing guidelines and standards to regulate the use of CFD for building design in early 2000s.

1.5 What Is CFD Used for?

Fluid mechanics governs all of the phenomena related to fluids and flows, from air to water to oil and from atmosphere to ocean to blood. As a promising means of fluid mechanics study, CFD has recently grown from a mathematical curiosity to an essential tool in almost every branch of fluid dynamics, ranging from aerospace propulsion to weather prediction. Today's CFD industry in the world is tightly coupled with various manufacturing and design industries such as automotive, aerospace, chemical and materials processing, power generation, biomedical, electronics, heating, ventilation and air conditioning (HVAC), generating over hundred billion profits every year. Professionals and non-professionals have been using personal computers and supercomputers to simulate flows in such diverse cases as the America's Cup racing yachts and blood movement through an artificial heart.

Knowing how fluids flow, and their quantitative effects on the solids with which they are in contact, can assist, for instance:

- architects and building engineers to provide comfortable and safe human environments;
- power-plant designers to attain maximum efficiency and reduce release of pollutants;
- chemical engineers to maximize the yields from their reactors and processing equipment;
- land/air/marine vehicle designers to achieve maximum performance at least cost;
- risk analysts and safety engineers to predict how much damage to structures, equipment, human beings, animals and vegetation will be caused by fires, explosions and blast waves.

In addition, CFD-based flow simulations enable:

- metropolitan authorities to determine optimal locations of pollutant-emitting industrial plants, and conditions that should restrict motor-vehicle access, to preserve air quality;
- meteorologists and oceanographers to foretell winds and water currents;
- hydrologists to forecast impacts of changes to ground-surface cover, of the creation of dams and aqueducts on the quantity and quality of water supplies;
- petroleum engineers to design optimum oil-recovery strategies and equipment systems;
- surgeons to understand probable consequences of potential surgery solutions on the flow of fluids within the human body (blood, urine, air, and/or the fluid within the brain).

In the last two decades, CFD has been playing an increasingly important role in building designs and environmental studies. The information provided by CFD can be used to analyze the impact of building exhausts to the environment, to predict smoke and fire risks in buildings, to quantify indoor environment quality, and to design natural ventilation systems, etc. The following paragraphs summarize the most important aspects in which CFD can assist in achieving a comfortable, healthy, and energy-efficient building design (Zhai 2006). The areas range from building site planning to individual room layout design, from active heating, ventilating and air-conditioning (HVAC) system design to passive ventilation study, and from regular indoor air quality assessment to critical fire smoke and contaminant control.

- **Application-1: site planning**

    Site planning is the first stage of building design. CFD can help optimize building sites by predicting the distributions of air velocity, temperature, moisture, turbulence intensity, and contaminant concentration around buildings. Good site planning can effectively protect building groups from adverse impacts of surrounding pollutions. It can also improve outdoor pedestrian comfort and increase energy efficiency of buildings by allowing passive HVAC strategies, such as using natural ventilation for summer and using wind break for winter.

Figure 1.4 presents such an example of using CFD for a site planning in Beijing, China. The initial plan introduces highly imbalanced airflow through the four high-rise buildings on the left of the figure, which may cause the pedestrian discomfort due to the large wind speed between some of the high-rise buildings. The non-uniform airflow pattern also reduces the chance of using natural ventilation for the buildings that confront less wind. The new design revised the building shape and orientation to allow natural wind to move smoothly cross each building so that it has a comfortable outdoor environment and has the same opportunity to use a natural ventilation strategy. Chen et al. (2000) indicated that natural ventilation can save about 40% of the total cooling energy required by buildings in Beijing.

Applying CFD for building site planning has become fairly convenient as most current commercial CFD programs can import AutoCAD files of building site models into the computational domain of a CFD simulation. The major remaining challenge is probably the long computing time due to the large number of mesh grids required to cover a building site with reasonable resolution. The computing cost may become more significant when dynamic wind conditions need to be modeled. Multi-grid and locally-refined grid technologies may, to some extent, accelerate the simulation; however, substantial computing time is still needed even with a multi-processor parallel computer.

- **Application-2: natural ventilation study**

Natural ventilation is one of the most fundamental techniques to reduce energy usage in buildings. In principle, CFD can simultaneously model indoor and outdoor airflows to achieve an optimal natural ventilation strategy. However, because of the scale difference between a typical room (meter) and a site plan (hundred meters), a large number of numerical grids must be used to meet the spatial resolution require-

(a) initial plan　　　　　　　　　　　　(b) final plan

Fig. 1.4 CFD for site planning

ment. This imposes an undue expense to designers by challenging current computer memory and speed. Therefore, a practical approach is to decouple the outdoor and indoor airflow simulation. Outdoor airflow around buildings is first predicted, which provides airflow and pressure information at the openings of buildings. With these boundary conditions, indoor airflow for each space can be simulated independently and natural ventilation rate can be determined. Designers can then change building indoor layouts and window sizes and locations to maximize natural ventilation rate.

The decoupled simulation method is based on the assumption that indoor airflow and building openings have little impact on outdoor airflow and pressure distributions; indoor and outdoor flow fields can therefore be studied separately. Zhai et al. (2000) verified that room partitions and windows often do not contribute to a major difference in outdoor flow patterns and pressure fields, if under a conventional window-wall-ratio. This decoupled method logically matches the general architectural design procedure: from site plan to unit design. The decoupled method first studies the outdoor airflow around solid building site models during the site plan stage when most details about building units are not determined yet; then it moves into building interior layout and opening design when the site plan is generally finalized. As a result, refining the microscopic unit design during the second stage of building design does not require the recalculation of the macroscopic site plan. Hence, the method greatly reduces the unnecessary computing time for dynamic design modifications, which allows designers to more easily refine the site plan and apartment and window layouts separately.

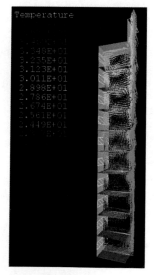

(a) buoyancy-driven natural ventilation   (b) wind-driven natural ventilation

Fig. 1.5 CFD for natural ventilation study

As examples, Fig. 1.5 shows the simulation results of (a) buoyancy-driven natural ventilation in a high-rise building with atrium and chimney, and (b) wind-driven natural ventilation in a typical one-floor apartment. These results provide designers with a straightforward understanding of the performance of natural ventilation design, and thus allow them to refine the plans and reach an optimal solution.

One of the major challenges of using CFD for natural ventilation study is the method to extract airflow conditions at building openings from outdoor simulations and specify them for indoor simulations. Because of the high sensitivity of CFD results to boundary conditions, small change of airflow conditions at openings may result in significant shift of indoor airflow patterns. In addition, simplification methods of indoor heat sources (e.g., occupant, equipment, etc.) also challenge indoor environment modeling.

- **Application-3: HVAC system design**

CFD is a powerful tool to evaluate indoor air quality and thermal comfort provided by diverse HVAC systems, leading to an effective and efficient system design. It is superior to the conventional design approach that typically relies on the use of charts provided by diffuser manufacturers and jet formulae that were developed from laboratory data. The use of such empirical data can result in great uncertainties when they are applied to large spaces (such as atria, concert halls and sports facilities) or applications that are dissimilar from those upon which the laboratory data were developed. When an innovative HVAC system is used, there are even no data or formula available for the engineering design.

Figure 1.6 illustrates the modeling of an office with innovative displacement ventilation systems. Displacement ventilation is an advanced indoor ventilation approach. Unlike the conventional mixing ventilation, displacement ventilation provides a cleaner indoor environment with less energy consumption. A typical displacement ventilation system supplies fresh air at or near floor level at a very low velocity and a temperature slightly below room temperature. Exhausts are located at or near the ceiling. The supply air spreads across the floor and rises as it is heated by sources such as people and equipment, removing indoor heat and contaminants directly from the occupied zone to the upper zone without mixing. Since only the occupied zone must be maintained at the room set-point temperature while the upper zone may be warmer, the supply air flow rate can be significantly reduced due to the vertical temperature gradient, resulting in the reduced fan energy. The CFD results help to understand the physics of the displacement ventilation (such as the large recirculation at the lower part of the room). They also quantify the vertical temperature stratification that is necessary for building energy calculation. Moreover, the supply air conditions can be optimized in CFD to reach the best comfort for occupants.

CFD can also be directly used to guide design process and optimize ventilation system design. Figures 1.7 and 1.8 demonstrate an example that uses CFD to design HVAC systems for the world's first large-scale indoor auto-racing facility. The facility is primarily a single space building with a floor area of over 0.2×10^6 m^2 and a ceiling height of 46 m. It is designed to accommodate up to 120,000 spectators—60,000 in the grandstands and 60,000 in the infield, as well as a maximum of 45 racing cars

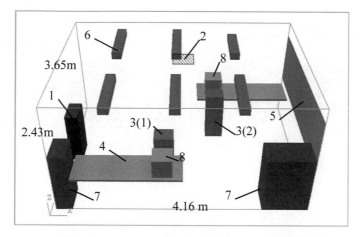

(a) inlet-1, outlet-2, person-3, table-4, window-5,
fluorescent lamps-6, cabinet-7, computer-8

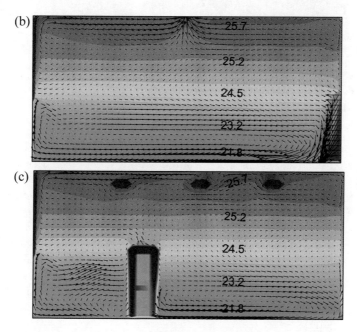

Fig. 1.6 Simulation of displacement ventilation in an office: **a** CFD model, inlet-1, outlet-2, person-3, table-4, window-5, fluorescent lamps-6, cabinet-7, computer-8; **b** velocity and temperature distribution in the middle plane of the room; **c** velocity and temperature distribution in the plane across the occupant 3(2)

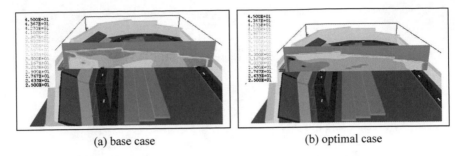

(a) base case   (b) optimal case

Fig. 1.7 Air temperature distribution in the middle section of an indoor auto-racing complex (unit: °C)

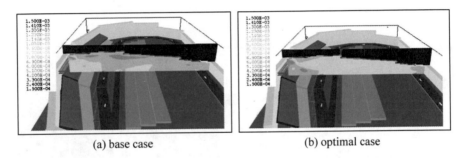

(a) base case   (b) optimal case

Fig. 1.8 Lead concentration distribution in the middle section of an indoor auto-racing complex (unit: gLead/kgAir)

running simultaneously on the track at an average speed of 217 km/h (135 mph). Such a large-scale and complicated building with a variety of indoor components strongly challenges the experience and capability of ventilation system designers, even with the aid of CFD modeling tools. CFD simulation has been used to improve the initial HVAC system design, step by step, to an optimal design. Figures 1.7 and 1.8 compare the air temperature and lead concentration in the mid-section of the facility under the steady design conditions by using different HVAC systems. The study concluded that a combination of underneath displacement ventilation system and conventional overhead duct system as well as partial air curtain system between occupied zone and racing zone is the most effective solution for this complex to obtain a comfortable and healthy indoor environment with less energy consumption (Zhai et al. 2002).

CFD results are much more informative and accurate than those that could be obtained via empirical, formulae-based hand calculations for HVAC design. However, CFD for HVAC system design still confronts various challenges, especially in the simplification of sophisticated building system components, such as diffusers, fans, evaporators, and diverse heat and contaminant sources (e.g., moving cars, breathing occupants).

- **Application-4: pollution dispersion and control**

CFD has widely demonstrated its capability in modeling the transportation of contaminants, with its low costs, high efficiency and flexibility. It is particularly useful for predictive studies in extreme conditions, and it can be easily employed to investigate the impact of a particular flow parameter, such as wind speed or air temperature, on the dispersion of a certain contaminant. Both indoor and outdoor contaminant dispersions can be simulated, while indoor scenarios are more complicated and hazardous because people spend over 90% of their time indoors and more factors can affect the dispersion of indoor contaminants. The geometry and structure of a building, as well as the HVAC system used in the building, have a dominant influence on the dispersion of indoor contaminants. Partitions, furniture and passageways between indoor spaces can also distort the airflow and the contaminant distributions. Importance of indoor contaminant study is also reflected by the fact that the indoor pollution is controllable by using good sensor and response systems. CFD prediction can be used to locate the best sensor positions in a building, to indicate the safe paths for evacuating occupants, and to develop the effective emergency response strategies to isolate and clean the contaminated air.

Figure 1.9 presents a realistic office complex as an example, on which CFD has been used to predict the dispersion of contaminants from different locations in the offices. The study showed that the contaminant dispersion is very fast and strongly depends on the indoor airflow pattern. It also indicated that early warning from the sensors is possible if they are placed properly. The investigation proposed and tested several response strategies by supplying or exhausting emergency air through three ceiling-mounted air devices. It found that the contaminant dispersion can be effectively controlled by simply pressurizing or vacuuming the indoor spaces.

Figure 1.10 illustrates another example of using CFD to design exhaust hoods for chemical and biological laboratories. The simulated results showed that, without special design cares, hoods with standard/enhanced ventilation rate may still leak toxic materials from operating zone to occupied zone due to the local turbulent vortices at hood openings. Hence, central air system and hood air system should be designed as a comprehensive system.

In general, the study of indoor contamination is not an easy task. The main technical concern is the development of appropriate physical and mathematical models to describe various contaminants with different phases and properties and to describe the interactions of contaminants with different objects (e.g., solid/soft surfaces, occupants, etc.).

1.6 How Does CFD Make Predictions?

CFD uses a computer to solve the relevant science-based mathematical equations, using information about the circumstances in question. Its components are therefore:

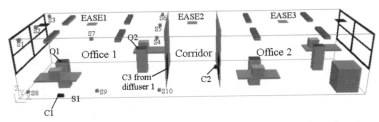

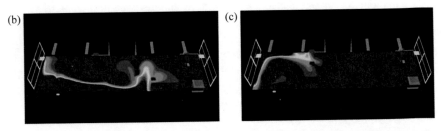

(a) C1~C3 represent three different types of airborne contaminants from three locations – under a desk in office 1 (C1), in the corridor (C2), and from the supply air in office 1 (C3). O1 and O2 are two occupants' nose locations (0.9 m above the floor) and S1~S10 are ten different sensor locations to be tested in office 1. EASE1~3 stand for three emergency air supply and exhaust (EASE) outlets

Fig. 1.9 Simulation of indoor contaminant dispersion and control: **a** CFD model; C1~C3 represent three different types of airborne contaminants from three locations—under a desk in office 1 (C1), in the corridor (C2), and from the supply air in office 1 (C3). O1 and O2 are two occupants' nose locations (0.9 m above the floor) and S1~S10 are ten different sensor locations to be tested in office 1. EASE1~3 stand for three emergency air supply and exhaust (EASE) outlets; **b** concentration contour of C1 at occupant head level at t = 5 min after contaminant release (without emergency response); **c** concentration contour of C1 at occupant head level at t = 5 min after contaminant release (with air pressurizing for corridor and office 2 and vacuuming for office 1 starting from t = 2 min)

- the human being who states the problem,
- scientific knowledge expressed mathematically,
- the computer code (i.e., software) which embodies this knowledge and expresses the stated problem in scientific terms,
- the computer hardware which performs the calculations dictated by the software, and
- the human being who inspects and interprets their results.

A successful CFD investigation requires the operator not only having strong computer and numeric skills but also holding profound fluid mechanics knowledge and rich hands-on experience in CFD. Scientific insight, knowledge, ability, and experience of CFD performers will determine whether one could correctly state and simplify the problem in both physics and mathematics, and whether one could correctly operate and improve CFD hardware and software, and whether one could correctly review, analyze and interpret CFD results.

(a)

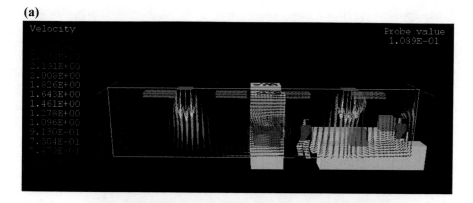

(b)

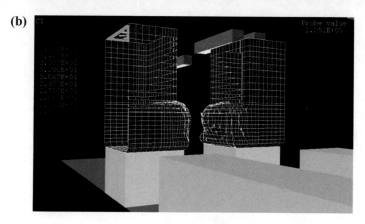

Fig. 1.10 Ventilation efficiency of hoods in chemical and biological laboratories. **a** Air velocity in the middle section of a lab. **b** Iso-surfaces of C = 1.052 ppm

Reality applications are so complicated and diverse that no single current CFD program can read them directly (probably not even viable in the next 20 years at least, if any possibility). Customized interpretation and approximation will be inevitable to translate the reality into a physical and mathematical model that can be understood and handled by a computer and computing code. Due to these approximations and assumptions adopted during the process, CFD results won't be valid (or trustable) until significant validations and verifications are conducted. A rigorous validation and verification procedure will be introduced later in this book.

1.7   Can CFD Be Trusted?

Similar to any other modeling and simulation approach, CFD-based predictions are never one hundred percent reliable and accurate. This is because:

- the input data may involve too much guess-work or imprecision. Examples include the assumed uniformity of inlet velocity and surface temperature, simplified geometries (from the actual objects such as human body), incomplete dynamic characteristics of boundary conditions in transient scenarios.
- the available computer power may be too small for high numerical accuracy. This may include insufficient computational grid and domain, or not small-enough time step and adequate iteration in modeling transient scenarios. Sometimes, it can be merely due to the requirement of fast turn-around of simulation results (e.g., for design purpose).
- the scientific knowledge base may be inadequate. This is especially true when new materials, systems, and processes are to be modeled, where their fundamentals (and thus physics and mathematics) are not well understood. Some special cases are those involved with mass transfer, phase changes and multi-phase interactions, such as, evaporation, condensation, boiling, and combustion.

In general, the reliability of CFD prediction is greater if a simpler case can be created and modeled, without compromising the fundamental mechanisms of the flow, where the flow knowledge is better understood in both physics and mathematics. The simple rules of thumb are:

- for laminar flows rather than turbulent ones, unless necessary;
- for single-phase flows rather than multi-phase flows, unless necessary;
- for chemically-inert rather than chemically-reactive materials, unless necessary;
- for single chemical reactions rather than multiple ones, unless necessary;
- for simple fluids rather than those of complex composition, unless necessary.

The bottom line is, although many actual situations can be very challenging to model (such as coal-fired furnaces) with large uncertainties, the use of CFD nevertheless provides more confidence with scientific grounds in designing and improving the complex realities than those obtained purely from knowledge or experience based guesses.

1.8   How Does One Become a CFD User and a CFD Pro?

Conventionally, three approaches are available to acquire the benefits of CFD.

Developing a general or dedicated in-house CFD code and becoming a CFD expert during this programming process was a dominant path when commercial codes were hardly available or too expensive to purchase. Another need to develop own CFD code is that the general commercial programs cannot perform special computing tasks due to the lack of simulation capabilities (such as to model phase change materials) and the lack of transparence of codes (so new models or functions created cannot be implemented into the code for investigation). Such exploration studies often occur in research entities where continuous efforts to develop, improve, and maintain the code may be possible. For most industries, developing an in-house CFD code would take much significant manpower (and thus cost) to undergo the entire programming, debugging, validation, and quality assurance process before the results can be presented and accepted with adequate confidence.

Purchasing a suitable computer code and learning how to use it then appears a wiser alternative. This becomes especially true when the numbers of CFD vendors and users are increasing and the license fee is thus reduced greatly. One may learn a CFD program by reading books and manuals, attending workshops offered by vendors, or taking a complete CFD course. Although the purchase of CFD software becomes more affordable, hiring experienced CFD scientists/engineers is still quite costly. Consequently, this approach may merely work, financially, for businesses with constant CFD project demands.

For businesses with occasional needs for CFD simulations, hiring a consultant to carry out the simulation and provide results and analysis will be much efficient (in both time and cost). Well-selected CFD consultants, who are educated and working in the CFD field for years and who even have worked on the same type of projects before, can start the project promptly and effectively with less warm-up time and back-and-forth processes. In practice, some of CFD consulting projects, depending on the project scale, may be supported by a team of CFD professionals. Results obtained by reputed CFD consults (or teams) could be relatively easier to be accepted by reviewers or clients.

The fundamental difference between a CFD user and a CFD professional is that a CFD professional knows well what is behind each button on the GUIs of a CFD program. Most user manuals and short-term workshops focus on the procedure of using the program, rather than digging into mathematical and numerical grounds, due to the constraints of space and time. This provides a potential challenge of examining, interpreting, and improving the simulations. A good GUI can dramatically reduce the input efforts and errors, which will attract more designers and engineers to apply CFD programs for their projects. But this may also jeopardize the credential of CFD, due to various faulty and misleading CFD results. Indeed, simple and intelligent GUIs should minimize the need to understand the underlying flow physics and numerical methods but still ensure obtaining meaningful results. However, this still does not address the key task in a CFD study—finding the problems and resolving the problems via the assistance of CFD. This requires profound knowledge in fluid dynamics and numerical science—the key for a great CFD pro. To this sense, a systematic learning of CFD principles and practices will be necessary if one is to become a CFD professional, rather than just a user. These principles and practices are commonly adopted by, and thus independent of, any CFD programs, either in-house or commercially available one. One, who completed this study, is anticipated to be able to use any CFD software after a short period of acquaintance with the GUI of that particular CFD program.

Practice-1: Outdoor Isothermal Flow

Example Project: Influences of Building Outdoor Landscaping on Natural Ventilation Design

Background:

Natural ventilation is an effective and energy-efficient method to remove extra indoor heat and moisture when outdoor conditions are comfortable. It is most feasible for residential buildings where the central air conditioning systems are less utilized. Chen et al. (2000) estimated that natural ventilation can save about 40% of the total cooling energy required by residential buildings in Beijing. The effectiveness of natural ventilation is influenced by both building site planning and indoor layout design (Zhai et al. 2000). Inappropriate building outdoor landscaping and indoor layout design may reduce the anticipated natural ventilation benefits. This study is to investigate the extra wind-resistance caused by outdoor trees. The study will employ a commercial computational fluid dynamics (CFD) program to simulate the natural ventilation rate change under different outdoor conditions.

To explore the wind-blocking effect of outdoor trees, a typical two-story building is simulated with a tree hedge located at the windward side of the building. The study computes and compares the pressure differences between the front and back façade of the building with trees at different locations. The pressure difference determines the potential natural ventilation rate through the building. The simulation results reveal the wind-blocking effect of outdoor trees at different distances from a building, which is a con for summer natural ventilation design but a pro for winter wind-break design (Zhai and Liu 2007).

Simulation Details:

For the outdoor landscaping study, the simulation predicts the steady-state air pressure and airflow field around a typical two-story building of $L \times W \times H = 16\,m \times 8\,m \times 8\,m$ by solving the Reynolds-Averaged Navier-Stokes (RANS) equations. The standard k-ε turbulence model (Launder and Spalding 1974) is used to represent the overall turbulence effect on airflow. Figure 1.11 illustrates the overall computational domain as well as the boundary conditions used. The whole domain is divided into about 600,000 grids with fine grids around the building and the trees. A wind velocity profile of $V = 7\,m/s \times \left(\text{Height}/10\,m\right)^{0.25}$ is employed to model the natural wind above the ground. The tree hedge is placed at the windward of the building, with the same length (L) and height (H) and half of the width (W) as the building. The trees have a typical leaf density and friction coefficient as observed from the field (Yi et al. 2005). The distance between the trees and the building was set at 0.5H, 1H, 2H, 3H, respectively. For each case, the study predicts the pressure distribution and airflow pattern around the trees and the building. The computed pressure differences between the front and back façade of the building for these cases are compared to the result of the reference case without the tree hedge.

X-Y Plane

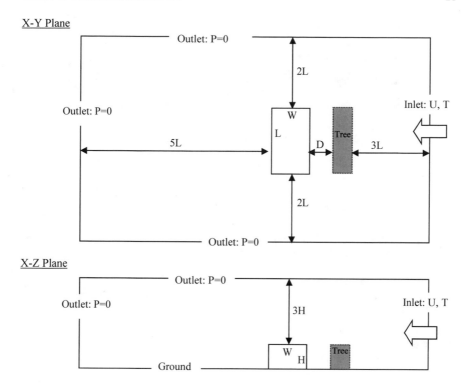

X-Z Plane

Fig. 1.11 Computational domain and boundary conditions (not scaled)

Results and Analysis:

Figure 1.12 shows the horizontal pressure contour plots at the middle plane of the building without and with trees. As anticipated, high pressure is observed on the windward side of the building, and low pressure on the leeward side of the building. The existence of trees results in an additional pressure drop through the trees, reducing the wind speed encountered by the building. Figure 1.13 presents the average pressure differences between the front and back facades of the building with

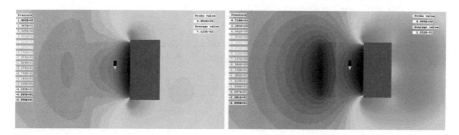

Fig. 1.12 Predicted pressure contours at the middle plane (XY) of the building (Z = 4 m) without trees (left) and with trees at D = 8 m from the building (right)

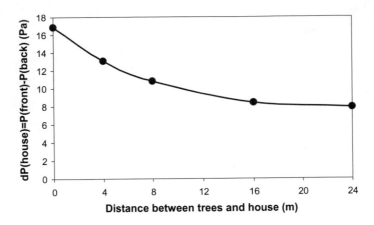

Fig. 1.13 Predicted pressure difference between the front and back façade of the building with different tree locations (D = 0 m means no trees)

different tree locations, in which higher pressure difference implies larger natural ventilation rate. It is evident that the natural ventilation rate decreases as the trees move away from the building. This seems somehow opposite to the human common sense that trees closer to buildings may block more natural wind. In fact, this can be well understood with the tree pressure reduction equation:

$$\Delta P_{drop} = CV^2_{local\text{-}air} \tag{1.1}$$

where C is the tree friction coefficient and $V_{local\text{-}air}$ is the local air speed in the tree. As seen in Fig. 1.14, the wind speed encounters gradual reduction when it approaches to the front façade of the building, particularly in the range of 16 m from the building front façade. As a result, the trees planted at a distance of 16 m from the building provide an almost full pressure drop based on the free incoming airflow velocity, which has little difference from the results of the trees at 24 m. When the trees approach to the building, the friction effect of trees is decreasing because of the decreasing local wind speed. One extreme situation is that trees planted on the front façade will have no friction effect on the wind because of the stagnant/zero wind speed on the façade. This indicates that when planning a site and the location of landscaping, trees that are meant to block the wind are probably most effective when placed at a certain distance from the building, versus immediately adjacent to it. Certainly, the influence of trees on building will decrease when the trees move too far away from the building because the airflow will be revived between the trees and the building.

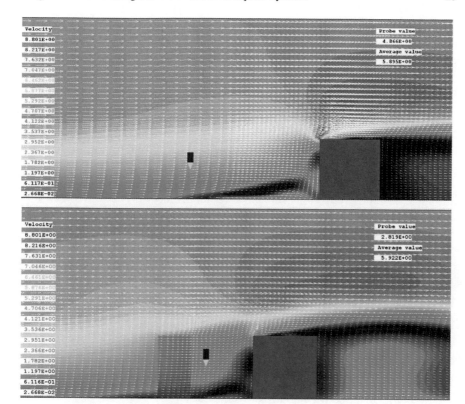

Fig. 1.14 Predicted velocity vectors and contours at the middle section (XZ) across the building without trees (top) and with trees at D = 8 m from the building (bottom)

Assignment-1: Simulating 2-D Flow Past a 2-D Square Cylinder

Objectives:

This assignment will use a computational fluid dynamics (CFD) program to model the external cross flow over a 2-D object.

Key learning point:

- External simulation with appropriate domain sizes
- Boundary condition settings
- Basic result analysis.

Simulation Steps:

(1) Build a square shape (L = 1 unit) at the center of the flow domain;
(2) Select appropriate flow domain sizes to be modeled;

(3) Prescribe proper boundary conditions including inflow (given flow conditions U), outflow (given P_{atm} or assumed fully developed flow), and both sides (given P_{atm} or assumed symmetric) [iso-thermal case only: no temperature];

(4) Calculate the Reynolds number: $Re = UL/\upsilon$;

(5) Select a turbulence model: the standard k-ε model;

(6) Define convergence criterion: 0.1%;

(7) Set iteration: at least 1000 steps for steady simulation;

(8) Determine proper grid resolution with local refinement: at least 200,000 cells.

Cases to Be Simulated:

(1) Test different domain sizes ($Re = 10^5$);

(2) Test three different inflow velocities U with distinct Re values ($Re = 10^4$, 10^5, 10^6) using the most suitable domain sizes.

Report:

(1) Case descriptions: description of the cases

(2) Simulation details: computational domain, grid cells, convergence status

  • Figure of the grids used (on X-Y plane);
  • Figure of simulation convergence records.

(3) Result and analysis

  • Figure of flow vectors across the body;
  • Figure of pressure contours across the body;
  • Figure of velocity contours across the body;
  • Compare the pressure distributions at the four surfaces of the square against the experiment results for the $Re = 10^5$ case (e.g., Ohtsuki 1978);
  • Evaluate the influences of domain sizes on simulation;
  • Compare the flow characterizes among the three cases (e.g., pressure distributions at surfaces, flow separation/vortex sizes around the body);

(4) Conclusions (findings, result implications, CFD experience and lessons, etc.)

References

Chen Q, Allocca C, Hamilton S, Huang J, Jiang Y, Kobayashi N, Zhai Z (2000) Building natural ventilation design and studies. In: Proceeding of conference on thermal environment of residential buildings, Beijing, China

Launder BE, Spalding DB (1974) The numerical computation of turbulent flows. Comput Methods Appl Mech Eng 3:269–289

Ohtsuki Y et al (1978) Wind tunnel experiments on aerodynamic forces and pressure distributions of rectangular cylinders in a uniform flow. In: Proceedings of the 5th symposium on wind effects on structures, Tokyo Japan, pp 169–175

Yi C et al (2005) Modeling and measuring the nighttime drainage flow in a high-elevation, subalpine forest with complex terrain. J Geophys Res Atmos 110

Zhai Z (2006) Applications of CFD in building design: aspects and trends. Indoor Built Environ 15(4):305–313

Zhai Z, Chen Q, Scanlon PW (2002) Design of ventilation system for an indoor auto racing complex. ASHRAE Trans 108(1):989–998

Zhai Z, Hamilton SD, Huang J, Allocca C, Kobayashi N, Chen Q (2000) Integration of indoor and outdoor airflow study for natural ventilation design using CFD. In: Proceedings of the 21st AIVC annual conference on innovations in ventilation technology, The Hague, The Netherlands

Zhai Z, Liu J (2007) Influences of building outdoor landscaping and indoor layout on natural ventilation design. J Harbin Inst Technol 14(Sup):318–321

Chapter 2
Model Real Problems

2.1 General CFD Modeling Procedure

Most commercial CFD programs now come along with powerful user-friendly graphic interfaces and detailed user manuals; however, proper and efficient usage of these programs still requires sufficient expertise in fluid mechanics and its numerical methods. The first crucial element in CFD simulation is a procedure or method of abstracting, simplifying, and reconstructing the real world into a computer model, which directly determine the correctness and accuracy of final predictive solutions. The knowledge and experience on similar problems can expedite this process. Furthermore, because CFD solves the non-linear partial differential equations for fluid flows, the performance of CFD prediction heavily depends on specific case conditions and characteristics. No general and simple rules can guarantee the convergence and stability of the solution. A professional with adequate knowledge of fluid physics and numerical techniques is always desired to solve complex flow problems. Lastly, the judgment and analysis of the results provided by computer simulation is also not a trivial job for people without appropriate expertise on fluid mechanics.

The following steps describe the general CFD modeling procedure:

(1) **Physical model simplification and setup**: Real environments (e.g., urban, building, or aircraft) to be studied are often complicated with numerous details that may have no or minor impact on the flow characteristics of interest. It is highly crucial to develop a reasonable CFD model with sufficient details that are elemental to the disclosure of flow physics. Other minor and overwhelming information not only provides no extra influence on the flow, but also imposes additional challenges and burdens to the simulation (e.g., grid distribution, numerical stability, convergence speed, etc.). Creating a successful CFD model requires some a priori analysis and insight on key flow physics in study, as well as clarification and understanding of primary modeling objectives that may determine the detail level to be included in the model.

(2) **Problem description**: This step specifies the flow characteristics and variables to be simulated (e.g., steady or unsteady flow, laminar or turbulent flow, isother-

© Springer Nature Singapore Pte Ltd. 2020
Z. Zhai, *Computational Fluid Dynamics for Built
and Natural Environments*, https://doi.org/10.1007/978-981-32-9820-0_2

mal or buoyancy flow, extra variable to be modeled such as water vapor or contaminant, etc.) based on prior knowledge of the flow as well as specific simulation goals and interests.

(3) **Physics property specification**: This step specifies the fluid properties, such as type of fluid and corresponding fluid properties (e.g., density, specific heat, viscosity, etc.)

(4) **Boundary and initial condition setup**: This is one of the most important steps in CFD, which simplifies and setups boundary conditions that enclose the flow domain. Unique boundary conditions distinguish various CFD scenarios even with the same built CFD model. For an unsteady flow, initial condition is also required to present the starting point (flow field) of a transient fluid flow. Proper acquisition, simplification and approximation of actual boundary (such as supply inlet and exhaust outlet) and initial conditions are always challenging but critical with significant consequence on simulation accuracy.

(5) **Discretization and grid generation**: This step discretizes the continuous spatial (and temporal) domain in the built CFD model into a large number of discrete small cells (and time steps for unsteady flow). Each of these discrete cells shares the same fluid and flow properties (e.g., density, viscosity, velocity, temperature, turbulence intensity, contaminant concentration etc.). A good grid/mesh distribution should have a sufficiently large grid number that can provide adequate resolution to reveal the flow details as well as delivering a grid-independent solution (to be discussed more in Chap. 9). A large grid number (i.e., fine spatial resolution) may provide great flow details but impose great challenges on computing power, computational speed, as well as numerical stability and convergence. In addition, the extra details provided by fine grids sometimes may not deliver additional useful information for result analysis. A reasonable balance between grid resolution and computational cost is always desired even with the rapid development of computer power. Furthermore, a proper grid system needs to ensure the good quality of cells generated, such as providing sufficient coverage at the flow fields with significant changes (or gradients) of variables (e.g., velocity next to solid surfaces), and avoiding sudden change of grid size between two adjacent cells that may result in additional numerical error and instability issues.

(6) **Numerical control parameter specification**: This step specifies various control parameters for numerical algorithms (such as relaxing factors that can help prevent simulation divergence). Today's commercial software usually implements various intelligent and automated controls of these parameters to relieve the dependence on users' involved knowledge and experience on these topics and thus reduces potential simulation failures. However, a user still needs to provide CFD simulation the criteria for stop or convergence, by either prescribed iteration number or allowed maximum residues in conservations of mass, momentum, and energy or both.

(7) **Run CFD engine**: A user can then trigger the CFD engine after all (1)–(6) steps are accomplished. Most CFD software (either commercial or in-house) will plot some intermediate results on a computer screen to visualize the real-time

performance of a simulation. These intermediate results may include predicted values of key variables at prescribed critical locations in the flow domain and/or accumulated residues in conservations of mass, momentum, and energy over the flow domain. The intermediate results can help determine whether the simulation is moving along a proper trend so that wrongful simulation can be terminated in time for re-evaluation.

(8) **Result analysis**: CFD results can be plotted in various formats (e.g., vector, contour, line profile, animation, etc.) and undergo further analysis. When a CFD solution is not successfully obtained (e.g., diverged), the following influential factors or checking steps are suggested for reviewing and refining the CFD case in order to ensure the solution's correctness, stability and converging-speed.

(a) Problem simplification: are they reasonably representing the primary physics and mechanics? Is the model too simple or too complicated?

(b) Computational domain: is it reasonably set or large enough to cover the flow field?

(c) Grid number and quality: is the resolution adequate and are they in good quality?

(d) Boundary conditions: are they reflecting the actual physics?

(e) Initial values: are they obtained or assumed correctly?

(f) Numerical scheme selection: are they appropriate or sufficient for this problem?

(g) Turbulence model selection: are they appropriate for this problem?

(h) Other numerical efforts (relaxation, iteration etc.): are they properly set for this problem?

2.2 General Rules for Model Simplification

Abstracting, simplifying, and reconstructing the real world into a computable (or computer-recognizable) model is always a great challenge, for any simulation task and tool. This requires great insight and prior experience on the physics and mechanisms of the problem in study. Two general rules should be emphasized during the model simplification process.

(1) **Simpler is better**. Whenever possible, a simple model (in geometry, physics, boundary condition, etc.) is always recommended, at least as a base case, which may readily produce a reasonable result. Advanced features can be added to this base model thereafter, step-by-step, to reach a comprehensive case configuration. In general, one should consider starting with

a. Inviscid flows rather than viscous ones;
b. laminar flows rather than turbulent ones;
c. Two-dimensional flows rather than three-dimensional ones;
d. Steady flows rather than unsteady or transient ones;

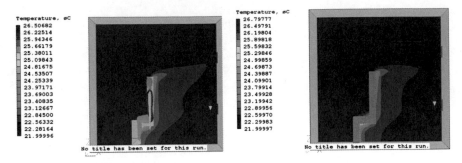

Fig. 2.1 Modeling of a sitting person with a realistic manikin and a blockage manikin

   e.   Incompressible flows rather than compressible ones;
   f.   Single-phase flows rather than multi-phase flows.

(2) **Focus on the main flow physics** of interest and the physical components and
     forces that contribute to the physics. Using geometry as an example, a curved
     surface may be important for a duct bend internal flow, but less critical for an
     urban environment modeling as the detail may not even be captured by numer-
     ical resolution (mesh) and may have less significant impact on primary flow
     characteristics of interest. Figure 2.1 shows the modeling of airflow across a
     sitting person, as another example, the person can be simplified as a blockage if
     the research focus is the general room air distribution, rather than the air around
     the body. A blockage person and a realistic person may have marginal differ-
     ence in producing general flow patterns. However, an over-simplified model
     (e.g., a solid blockage person ignoring the gap between two legs) may produce
     noticeable disparities in flow characteristics behind the legs, deviated from that
     considering the leg gap. On the contrary, if the micro-environment around the
     person is the target of the study, e.g., to investigate the inhale risk of contam-
     inants, a precise human geometry will be necessary. Indeed, studies show that
     the size of nose and chin will have an impact on the local airflow pattern (and
     thus contaminant dispersion).

2.3   Classification of Fluids and Flows

This session provides general knowledge of fluid and flow classifications, which can
assist one to choose and create appropriate CFD models. Particular attentions are paid
to distinguishing the actual and ideal (simplified) flow properties and conditions.

   Fluids and flows can be classified into several categories, depending on the properties and conditions of fluids and flows in study. Table 2.1 summarizes the common
properties used to classify various fluids and their flows. The categories in *Italic*

Table 2.1 Types of fluids and fluid flows commonly used to solve fluid flow problems

Property	Variable (V) or applicable (A)	Constant (C) or not applicable (NA)
Viscosity	Viscous flows (V)	*Inviscid (frictionless) flows* (C)
Density	Compressible fluids (V)	*Incompressible fluids* (C)
Time	Unsteady (transient) flows (V)	*Steady flows* (C)
Direction	2- or 3-Dimensional flows (V)	*One-Dimensional flows* (C)
Orderly flow	Laminar flows (A)	Turbulent flows (NA)
Initial force	Forced flows (A)	Natural flows (NA)
Inside flow	Internal flows (A)	External flows (NA)

represent simplifications or approximations of actual fluid properties and flow phenomena in physics, which can facilitate the analysis of fluid mechanics.

In physics, all fluids have viscosities (internal stickiness) due to the cohesive forces of liquid molecules or the collision of gas molecules. Viscosity is the cause of surface friction, which converts mechanical work into thermal energy. When the viscous (friction) force is much smaller than other forces (such as forces caused by pressure difference or body gravity), the viscous effort can be neglected, which is called inviscid flow. Assumption of invisicid flow is commonly used when studying high-speed aircraft where the friction caused by the pressure difference (due to the aircraft shape) is much larger than the surface friction.

Similarly, all fluids can be compressed under a certain pressure. However, when the density of a fluid is not changed significantly during a process, the fluid can be treated as incompressible. For air, typically, if the density variation during a process is less than 5% of the original air density, it can be handled as incompressible fluid, in which the air density is not the function of time. For air under typical room conditions (temperature and pressure), $\Delta\rho_{air} < 0.5\% \ \rho_{air}$ is equivalent to $V_{air} < 100$ m/s. Therefore, most room airflows (typically at 0.1–10 m/s) can be treated as incompressible flows.

In reality, all fluid flow properties (e.g., pressure, velocity, and temperature) are the function of location (spatial) and time (temporal), i.e., $\phi = \phi(x, y, z, t)$. However, if the variation of the properties is independent of some argument (e.g., z or t), the flow characteristics can be simplified in mathematics. It can, respectively, become steady flow, $\phi = \phi(x, y, z)$ if the temporal variation is small, two-dimensional flow, $\phi = \phi(x, y, t)$ if the variation in z direction is minor, or one-dimensional flow, $\phi = \phi(x, t)$ if the variation in both y and z is negligible.

The majority of fluid flows in nature are turbulent (chaotic molecule movement). However, turbulence has obtained its infamous reputation because dealing with it mathematically is one of the most notoriously thorny problems of classical physics. The study of turbulence, which is the rule not the exception in fluid dynamics, although started over 100 years ago, is still in the somewhere middle. The understanding, prediction and even control of turbulence would change the world revolutionarily, which allows the appearance of more exciting techniques and devices, such as, faster transportation vehicles, smarter artificial organs, and more accurate broadcast of weather. Laminar flow typically occurs at a slower velocity, or strictly speaking, at a low Reynolds number. In laminar flows, fluid molecules tend to move along the streamlines, which in turn leads to less friction and heat transfer. Both laminar and turbulent flows can be found in reality. The most challenging flow in physics is the flow that starts to shift from laminar to turbulence, which is called transitional flow. Well understanding and modeling of this physics is still under investigation.

Depending on the driving forces, the flows can be divided into forced flows and natural flows, where the former is driven by some mechanical forces (such as fan and wind) while the latter is caused by temperature or density gradient (named as buoyancy). In most real conditions, both mechanical and buoyancy forces may exist simultaneously. Understanding the primary driving forces of a flow can help determine appropriate equations and terms to be solved in CFD. For instance, for wind-driven urban airflow modeling where temperature gradient and heat sources are neglected, the buoyancy force can be ignored and the energy (or temperature) equation does not need be solved. Similarly, the actual flows can be internal flows (e.g., flow in pipes and ducts), external flows (e.g., flow around bridges and buildings), or combined flows (e.g., flow across a building with large openings). Determining whether internal or external flows may affect the proper choice of computational flow domain size and flow boundaries.

2.4   Definition of Computational Domain

Determining suitable computational domain is critical for achieving a physically-correct and numerically-converged simulation. Computational domain is part of the actual physical domain, which is deliberately chosen to represent the flow mechanisms of interest. Larger computational domain may increase the computing cost that is not necessary to contribute new information to the essential physics. On the contrary, smaller computational domain may distort the actual flows due to the imposed artificial boundary conditions.

(a) **Internal Flows or Flows in Confined Spaces**

For internal flows or flows in confined spaces, physical boundaries (e.g., walls, partitions, etc.) are typically used as computational domain boundaries, such as airflow modeling in the aircraft cabin as showed in Fig. 2.2.

(b) **External Flows or Flows around Objects**

For external flows or flows around objects, adequate computational domain is often required. This is due to the need to avoid the challenges of determining the required conditions at boundaries of computational domains. Figure 2.3 shows a cube cross flow as an example. If the dash line is chosen to be the back side of the domain, either the pressure or the velocity profiles need be given as the boundary conditions, which however are unknown because of the complex vortices at the back of the cube. The simple solution is to extend the back domain to the location where the

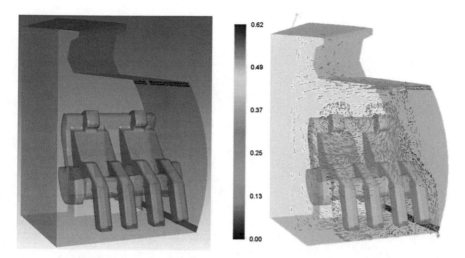

Fig. 2.2 Airflow velocity prediction in confined aircraft cabin

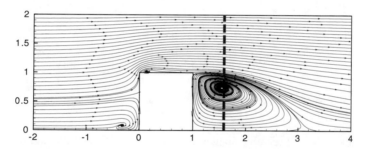

Fig. 2.3 Flow across a cube

flow is recovered from the disturbance of the cross flow and thus either atmospheric pressure or incoming flow profile can be assigned at the outflow. The similar principle is applied to find proper inlet, upper and side domain locations.

There is no firm conclusion for a right domain size as this may vary with flow cases and study interests. As a general guideline, the following rules of thumbs are suggested for engineering purpose (Fig. 2.4). For scientific researches, much large domains are often recommended, for instance, 3.5L for the upwind, 10L for down-wind, and 3L for both side-winds, to ensure the full development of both inflow and outflows.

(c) Coupled Internal and External Flows

For cases with coupled internal and external flows, a large domain often is needed that covers both internal and external flow fields and supplies with feasible boundary conditions for both. A large number of computational grids are usually required to cover the large external field and provide an adequate spatial resolution for the

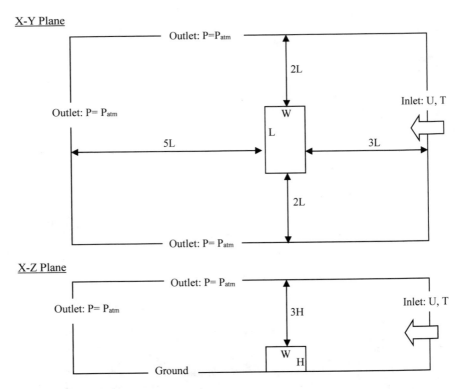

Fig. 2.4 Computational domain and boundary conditions for external flows (not scaled)

internal field. This is specially challenging when a significant scale difference exists between the internal domain (e.g., a typical room in meters) and the external domain (e.g., a site plan in hundred meters).

A practical alternative is to decouple the external and internal flow simulations. The airflow in and around objects such as buildings can be broken up into two entities: a macroscopic model for the general external flow around objects and a microscopic model for the flow details within the objects, e.g., individual buildings. The two airflows are reintegrated by exchanging the inter-connected boundary conditions. The external flow simulation provides the airflow and pressure information at the openings used in the internal flow simulations. Under these conditions, indoor airflow for each object can be studied independently.

This decoupled integration method is based on the assumption that indoor airflow and openings have little impact on the outdoor airflow and pressure distribution; indoor and outdoor flow fields can therefore be studied separately. This method completely severs any ties between the outdoor and indoor domains, except for the information extracted from the former and used in the latter. Hence, whether this method is workable is determined by whether the unit openings and interior layouts have significant influence on the airflow and pressure around the exterior of the object. By comparing the pressure information around a solid block building model and a corresponding hollow shell building model (openings constitute about 27% of the façade area) in a site plan, Zhai et al. (2000) found that the pressure differences between the windward and leeward sides of these two buildings have a design-acceptable cumulative error (less than 10%). It is thus reasonable to conclude that room partitions and windows do not contribute to a major difference in outdoor flow pattern and pressure field, and that values from outdoor simulation with solid block building models can be used as boundaries for indoor simulation.

Using a site plan and an apartment complex design as an example, the following paragraphs demonstrates the process of reintegrating the separate outdoor and indoor simulations for the incorporated design of natural ventilation. The evaluated apartment complex, which is 36-m high and consists of twelve stories and forty-eight apartments, is a part of the developing site that includes many different buildings, shown in Fig. 2.5. For the outdoor simulation, solid blockages were used to represent buildings in this site without considering openings in the building façades. Based on the weather data of the design location, the study assumed an average wind speed of 3 m/s (at 10 m above the ground) that originates from the southeast. To model the wind profile below 10 m, an exponential function $V = 3\,\text{m/s} \times \left(\text{Height}/10\,\text{m}\right)^{0.25}$ incorporates ground roughness in the boundary layer.

The decoupled approach first studied the pressure distribution on this site, with the aim to create the maximum pressure difference between the windward and leeward sides of the building designed. To achieve this goal, the location, shape and orientation of the building to be designed requires an iterative design, simulation, and redesign

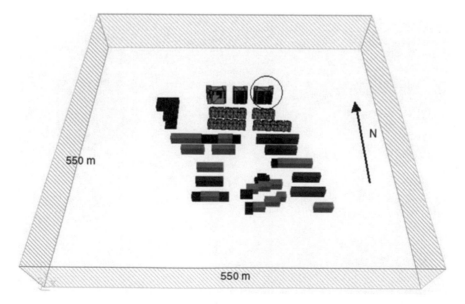

Fig. 2.5 Site plan where the apartment building studied is circled

process. Figures 2.6 shows a portion of the localized flow of the 550 m × 550 m site plan modeled. The windward and leeward pressure differences in the optimal design are about 4.7 Pa (South to North) and 6.1 Pa (East to West), which is favorable for natural ventilation design.

The pressure values at the building façade, halfway up the building, were then extracted to serve as the boundary conditions at the windows of those individual apartments to be designed. A typical duplex apartment and a single-level apartment in this building were studied. Figure 2.7 shows the locations and designed floor plans

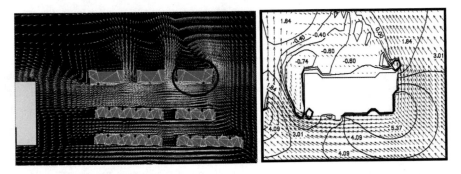

Fig. 2.6   Solid block site plan airflow field and pressure contour

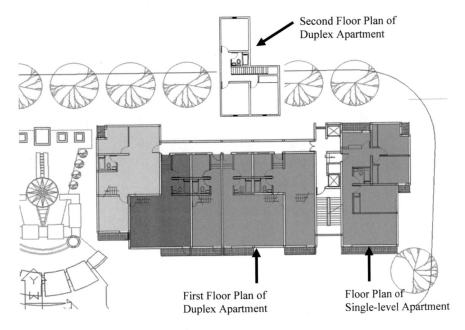

Fig. 2.7   Locations and floor plans of apartments studied in the building

of the apartments within the building. Furnishings are also added to both apartment units in order to determine how a typical interior configuration affects the airflow.

To obtain the maximum natural ventilation rate with the given wind pressure conditions, it is necessary to refine and optimize the apartment interior layout plan and window placement. Figures 2.8 shows the indoor airflow patterns (through a horizontal plane at the center height of the windows) in the optimal duplex and single-level apartments, where both of the airflows can smoothly enter the windward windows and exit the leeward windows of the apartments. The natural ventilation rate through each apartment can then be calculated, which is about 38 ACH and 19 ACH for the single-level apartment and the duplex apartment, respectively. The hand-calculation results of the airflow rate through these two apartments are, correspondingly, 27 ACH and 17 ACH, which converts pressure gradients into velocities by using a discharge coefficient of 0.6 through the windows and doorways of the apartments. It is interesting to note that the hand calculation method successfully predicts the ventilation rate for the duplex apartment, but not for the single level apartment. This is because the location of the single-level apartment at the corner of the building produces complex two-dimensional flow while the duplex (located in the center of the building) has roughly one-dimensional flow.

The above design practice reveals that the indoor and outdoor decoupled simulation method promotes predictive and iterative design solutions to optimize natural ventilation. The method allows the airflow simulation to be a practical design tool

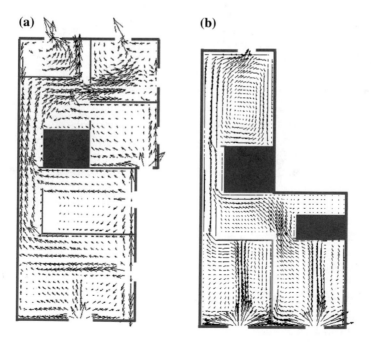

Fig. 2.8 Velocity vectors at the center height of windows for **a** the upper floor of the duplex apartment (the lower floor has similar results) and **b** the single-level apartment

that can work on a desktop computer with less computer power demand. The decoupled simulation method logically matches the general architectural design procedures well: from site plan to unit design. It first studies the outdoor airflow around the solid building models during the site plan stage when the most details about building units are not determined yet; then it moves into the building interior layout and opening designs when the site plan is generally finalized. As a result, refining the microscopic unit design during the second stage of building design does not require the recalculation of the macroscopic site plan. Therefore, the method largely reduces the unnecessary computing time for the dynamic design changes, which allows designers to more easily refine the site plan and apartment and window layout separately.

2.5 Abstraction of Physical Objects

Most physical objects to be modeled in CFD are sophisticated so that proper simplification and abstraction are necessary. As stated before, analyzing and focusing on main flow physics and influential factors is the primary principle for building abstract yet realistic CFD models. The following sessions use two actual design cases to demonstrate and illustrate the principles and practices of simplifying and approximating CFD models.

(1) Example-1: indoor auto-racing facility

The first case is to use CFD to optimize the ventilation system design for a large-scale indoor auto-racing facility. The facility is primarily a single space building with a floor area of over 0.2×10^6 m^2 and a ceiling height of 46 m, as shown in Fig. 2.9. The space is being designed to accommodate a variety of future possible occupancy conditions for a wide variety of events—60,000 spectators in the grandstands and/or 60,000 spectators in the infield, as well as lesser occupancies within various areas of the infield. The facility has special lighting and large screen displays for televised events, food and retail concessions stands, etc. The track facility is designed for a maximum of 45 racing cars running simultaneously on the track at a maximum speed of 242 km/h (150 mph) and an average speed of 217 km/h (135 mph). Such a large-scale and complicated building with a variety of indoor components strongly challenges the experience and capability of ventilation system designers, as well as CFD modelers.

The design is especially challenging because of the high speed of racing cars, the huge scale difference between the building and the internal objects (such as spectators), and the enormous amount of heat and chemical components generated from the fuel used by the cars. Special techniques and simplifications are needed to create the CFD model that translates the real world into a description of the flow physics suitable for numerical processing. The following paragraphs briefly describe the simplifications and techniques used in the study to handle the major thermo-fluid components in the complex.

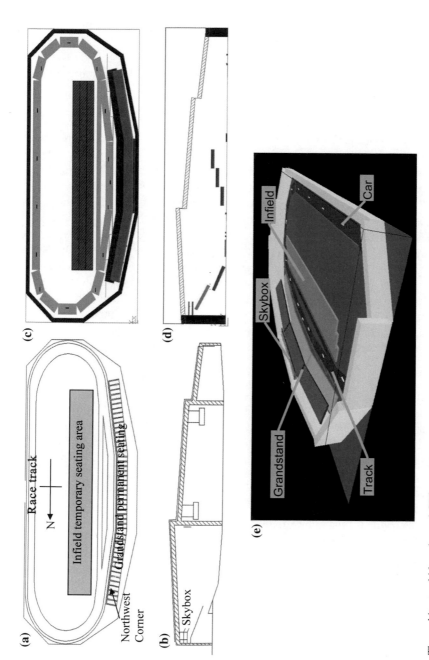

Fig. 2.9 The architectural blueprint and CFD model for the auto-racing complex: **a** blueprint plane, **b** blueprint middle section, **c** CFD model plane, **d** CFD model middle section, **e** 3D CFD model

(a) **Building Enclosure**

The CFD model constructed should represent the auto-racing facility in a similar but abstract manner, as shown in Fig. 2.9. The CFD models in Figs. 2.9c, d and the architectural models in Figs. 2.9a, b look alike, but there are differences. For example, the curved racetrack was simulated in the CFD model using square blocks. The local summer design conditions was used to estimate the interior surface temperatures of the walls and roofs.

(b) **Cars**

Simulating moving cars with moving boundary techniques for such a large space is almost impractical. The impact of moving cars on the indoor environment can be reasonably approximated by considering their velocity momentum and their affects as heat and contaminant sources. Therefore, the CFD model simulated these 45 cars as "still" objects with momentum, heat and contaminant source characteristics. This approach has been proved to be acceptable and practical in the study by Yang et al. (2000), which used this technique to simulate a moving ice resurfacer in an ice rink and obtained satisfactory results. A further simplifying technique used in this study was to group the 45 racing cars into 15 groups of three cars each, which represents common racing scenarios and reduces the input efforts. The 15 groups of cars were uniformly distributed on the track and assumed to be traveling at the same average speed of 217 km/h (135 mph) and with the same heat and contaminant generation rates (e.g., 750 horsepower or 599 kW per car and 3 kg/hour lead from the gasoline used by 45 cars within a typical three to five hours racing event).

(c) **Spectators**

The total heat generated by each person in such an event is around 150 W, and the moisture generation rate is 0.055 kg/h per person. However, due to the scale-difference and input-quantity limitations, it is impossible to simulate so many spectators individually in the CFD model. Therefore, by focusing on the macro influence of the spectators on the indoor environment, all of the spectators can be simplified into several solid blocks of resistance, heat and moisture sources, with an average occupied area of 0.5 m^2 for each person. The ventilation systems must be designed to work under a worst-case (or full-load) scenario. Hence, a major auto-racing event was modeled with the maximum number of spectators inside the complex and a maximum number of racing cars on the track, under the local summer design conditions.

(d) **Diffusers**

Since the size of diffusers is much smaller than those of other components in the building and also the types of diffusers have not been specified at the early design stage, the study employed the uniform air-supply assumption for all the diffusers with the usage of the momentum method (Chen and Moser 1991) assuming 50% actual supply area of the gross diffuser opening area.

With these simplifications, a base ventilation system design for this space was established, as illustrated in Fig. 2.10. The initial concept included supplying fresh

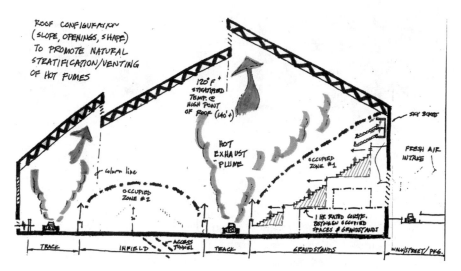

Fig. 2.10 Illustration of the ventilation strategy

air to the grandstands (occupied zone #1) and the infield (occupied zone #2); in addition, air-curtains between the track and the occupied zones were envisioned to help isolate the occupied zones from the hot and contaminated plumes generated by the cars. The rising hot and polluted air plumes were then to be mechanically exhausted from a series of large exhaust fans located along two clerestories at the roof level. The grandstand area of the base design assumed a traditional overhead duct system to supply fresh air, while the infield area was assumed to be ventilated by a displacement ventilation system that supplies fresh air underneath the seats. Rather than attempting to provide full air-conditioning of the entire facility space during a racing event, the design goal was to use the required ventilation air to provide partial "spot-cooling" of occupied areas, which would provide comfort levels similar to that experienced by racing fans in a conventional outdoor race track.

(1)   **Example-2: office complex**

The second case simulated a realistic office complex to predict chemical and biological agent (CBA) dispersion under a terrorist attack and to develop building-protection strategies (Zhai et al. 2003). The study modeled a typical 'linear' office building that has offices in both sides and a corridor in between. Assuming all the offices are identical, this study solely simulated a building section as shown in Fig. 2.11, to save the computing time and effort. Each office has two occupants, two tables, two computers, and four lamps, and one of them has a copy machine as summarized in Table 2.2. Since the study was focused on the primary air and contaminant distributions in the spaces, these objects were highly simplified and abstracted into the simple rectangular blockages with the same scales and thermal conditions as the actual objects (as seen in Fig. 2.11).

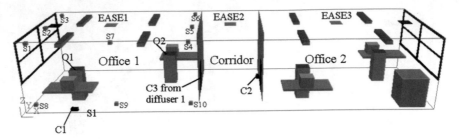

Fig. 2.11 Schematic of a section of the office building with various blockage objects

Assuming the two offices were in the middle of the building, the walls, ceiling and floor can be approximated as adiabatic surfaces. As a result, the sole thermal sources were the internal heat gains. The office complex were air-conditioned with a displacement ventilation system, which supplied fresh air from the lower part of the offices and exhausted contaminated air from the ceilings. Because of the air-conditioning, the higher indoor air pressure than that outdoors limited the outdoor CBA's penetration into the indoor space by infiltration. The space was assumed under several different types of indoor airborne CBA attacks from three locations—under a desk in office 1 (C1), in the corridor (C2), and from the supply air in office 1 (C3). Table 2.3 provides the locations and the release rates of these CBAs. The actual CBA sizes and properties were unclear and may vary with situations of interest. Since this study focused on a general fundamental research, and more importantly because the contaminant concentration has a linear relationship with flow velocity for steady-state flows, contaminant source sizes and concentrations have limited influence on simulated findings. A relatively large contaminant source size (10–20 cm) was used mostly to avoid fine grids around these objects and reduce the scale (and thus grid size) difference between these and other indoor objects. The research proved that these simplifications and approximations were appropriate and necessary for obtaining meaningful results with a reasonable computing budget.

Practice-2: Forest Canopy Flow

Example Project: Modeling Nighttime Drainage Flow in a High-Elevation, Sub-alpine Forest with Complex Terrain
   Background:
   The exchange of materials and energy between plant canopies and the atmosphere underlies some of the most important environmental challenges facing humankind, including perturbations to the global carbon cycle, the introduction of pollutants into the atmosphere and the transfer of water from soil and vegetation to the atmo-sphere. Land-atmosphere exchange is the key link between biosphere and atmo-sphere. During the past decade tower flux networks have flourished for monitoring

Table 2.2 Office complex configurations

Objects	Length	Width	Height	Location			Heat[a]
	Δx (m)	Δy (m)	Δz (m)	x (m)	y (m)	z (m)	Q (W)
Office 1	5.16	3.65	2.43	0.0	0.0	0.0	
Window 1	0.0	3.65	1.43	12.32	0.0	1.0	
Diffuser 1	0.0	0.65	1.0	5.16	1.5	0.0	
Exhaust 1	0.35	0.35	0.0	0.4	1.65	2.43	
Door Opening 1	0.0	0.75	2.43	5.16	0.0	0.0	
Occupant 11	0.40	0.35	1.1	1.1	0.95	0.0	75
Occupant 12	0.40	0.35	1.1	3.89	2.35	0.0	75
Computer 11	0.40	0.35	0.35	1.1	0.1	0.75	108.5
Computer 12	0.40	0.35	0.35	3.89	3.2	0.75	173.4
Table 11	1.47	0.75	0.05	0.58	0.0	0.7	
Table 12	1.47	0.75	0.05	3.69	2.9	0.7	
Lamp 11	0.2	1.2	0.1	1.1	0.1	2.23	68
Lamp 12	0.2	1.2	0.1	1.1	2.35	2.23	68
Lamp 13	0.2	1.2	0.1	3.49	0.1	2.23	68
Lamp 14	0.2	1.2	0.1	3.49	2.35	2.23	68
Office 2	5.16	3.65	2.43	7.16	0.0	0.0	
Window 2	0.0	3.65	1.43	12.32	0.0	1.0	
Diffuser 2	0.0	0.65	1.0	7.16	1.5	0.0	
Exhaust 2	0.35	0.35	0.0	11.57	1.65	2.43	
Door Opening 2	0.0	0.75	2.43	7.16	0.0	0.0	
Occupant 21	0.40	0.35	1.1	8.26	0.95	0.0	75
Occupant 22	0.40	0.35	1.1	11.05	2.35	0.0	75
Computer 21	0.40	0.35	0.35	8.26	0.1	0.75	108.5
Computer 22	0.40	0.35	0.35	11.05	3.2	0.75	173.4
Table 21	1.47	0.75	0.05	7.74	0.0	0.7	
Table 22	1.47	0.75	0.05	10.85	2.9	0.7	
Lamp 21	0.2	1.2	0.1	8.26	0.1	2.23	68
Lamp 22	0.2	1.2	0.1	8.26	2.35	2.23	68
Lamp 23	0.2	1.2	0.1	10.65	0.1	2.23	68
Lamp 24	0.2	1.2	0.1	10.65	2.35	2.23	68
Xerox machine	0.8	0.8	1.0	11.32	0.2	0.0	1600
Corridor	2.0	3.65	2.43	5.16	0.0	0.0	
EASE 1	0.35	0.35	0.0	2.20	1.65	2.43	
EASE 2	0.35	0.35	0.0	5.985	1.65	2.43	
EASE 3	0.35	0.35	0.0	9.38	1.65	2.43	

[a]*Note* The heat generated includes radiation and convection

Table 2.3 CBA locations and sources

Contaminant sources	Length Δx (m)	Width Δy (m)	Height Δz (m)	Location x (m)	y (m)	z (m)	Generation rate or concentration
C1 (in Office 1)	0.2	0.1	0.1	1.3	0.1	0.0	2.5 mg/s
C2 (in the corridor)	0.1	0.2	0.1	7.06	2.7	0.0	2.5 mg/s
C3 (from Diffuser 1)	0.0	0.65	1.0	5.16	1.5	0.0	2.5 mg/s or 42 ppm

land-atmosphere exchange by the eddy covariance approach. The Fluxnet tower network includes 368 sites as of October 2005. The eddy covariance approach is most accurate when applied to ecosystems with flat topography and homogeneous vegetation cover. However, many Fluxnet sites are located in complex terrain where topographic advection errors can be the same order as the eddy flux itself. Networks of flux towers are now vital to provide the empirical constraint required for accurate regional and global carbon budget modeling. However, advection caused by topography and surface heterogeneity remains a serious obstacle to routine 24 h operation for eddy flux towers. An international workshop (held in Boulder 26–28 January 2006) organized by the eddy flux research community defined advective flows as 'difficult conditions' in flux measurements.

This study reports significant progress in modeling canopy flows over complex terrain and in understanding how topographical flow influences CO_2 flux measurements (Yi et al. 2005).

Simulation Details:

To verify the experimental and analytical results, airflow within and above the canopy was simulated by using computational fluid dynamics (CFD) techniques. The CFD simulation solved the two-dimensional steady-state incompressible Reynolds-averaged Navier-Stokes equations of fluid flow. A gravity term $-g_i \eta (T - T_\infty)$ was included that represents the buoyancy force on fluid flow, where g_i is the gravity acceleration in i-direction, η is the thermal expansion coefficient of air, and T_∞ is the reference temperature. The drag force (pressure drop) F_D exerted by plant elements was also included in the Navier-Stokes equations.

$$F_D = \frac{1}{2} K_r u^2 \tag{2.1}$$

where K_r is the resistance coefficient, which can be related to the porosity by an empirical relationship given by Hoener (1965)

$$K_r = \frac{1}{2} \left[\frac{3}{2\beta} - 1 \right]^2 \tag{2.2}$$

where β is the porosity that can be determined from the leaf area density and drag coefficient profiles.

This study used a renormalized-group k-ε turbulence model (Yakhot et al. 1992), which generally has better accuracy and numerical stability than the "standard" k-ε model (Launder and Spalding 1974). The boundary conditions involved in this study include:

- Inflow: upwind wind profile specified with the semi-logarithmic law and $T = 7\,°C$.
- Outflow: leeward and sky with fixed static pressure of 1 atm.
- Ground: non-slip condition with negative heat flux of -20 W m^{-2}.
- Internal objects (plants) with resistance and heat: the resistance is specified according to Eq. (2.1) and the long-wave radiation from the plants is specified to linearly decrease from 63 W m^{-2} at the top layer of the canopy to zero at the middle of canopy [i.e., the stable situation specified in Siqueira and Katul (2002)].

The simulation divided the flow field into $200 \times 200 = 40{,}000$ cells in x-z section, which represents one computational node per 4 m in the x coordinate, per 0.15 m in the z coordinate within the canopy volume (0–16 m height), and per 0.56 m in the z coordinate above the canopy (17–28 m height). The computing time for such a simulation was about 4 h on a PIII-900 MHz desktop PC. Figure 2.12a illustrates the computational domain.

Results and Analysis:

Figure 2.12b shows the calculated velocity vectors and contours within and above the canopy. The secondary wind speed maximum due to drainage flow near ground and minimum wind speed near the canopy level with maximum leaf area density are clear in the CFD-simulated results. Caution should be taken in interpreting the simulated velocity field in the upper part of the domain (30–50 m) as the possible influence of larger-scale air motions like mountain waves (Durran 1990; Turnipseed et al. 2004) were not considered in this simulation.

Overall, the simulated wind profile within and above the canopy is in excellent agreement with the observations and analytical solution, especially below 10 m (Fig. 2.13) (Yi et al. 2005). Small differences between the results from the CFD model and observations above the 10 m height probably result from use of the semi-logarithmic law to smooth the observational wind profile. This would be consistent with past studies that have shown the local flux-gradient relationship (Raupach and Thom 1981) to be inadequate in the roughness sublayer.

Further CFD experiments found three different dynamic regimes of topographic drainage flow under different thermal-dynamic conditions (Fig. 2.14). Cold inflow induces drainage flow in the lower part of canopy and strong stratification of airflows within entire canopy; additionally, the model predicts that there is a super stable layer around the maximum leaf area density level, which is consistent with the canopy flow theory (Fig. 2.14a). This super stable layer minimizes vertical land-atmosphere exchange around the middle level of canopy. Warm inflow causes the rapid flushing of land-atmosphere exchange at the location where two opposite air motions meet, which is called the 'chimney phenomenon' (Fig. 2.14b). The oscillation of canopy flow occurs as the inflow temperature is close to the environmental temperature (Fig. 2.14c).

(a) Domain

Semi-logarithmic wind profile

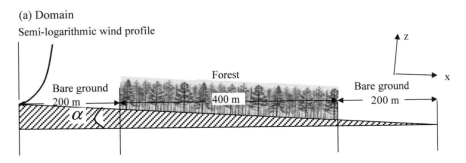

(b) Canopy flow

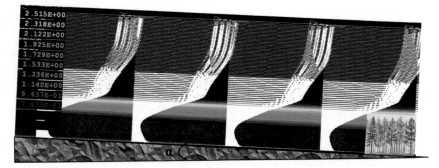

Fig. 2.12 **a** Domain and **b** The simulated two-dimensional canopy flow by the renormalized-group k-ε turbulence model. The drag coefficient profile derived from the analytical model and the leaf area density profile were used as inputs in the numerical model. In this simulation, slope $\alpha = 5$ degree, temperature deficit $\Delta T = -6\,°C$, and details for the boundary conditions are described in the text. The flow field shown in (**b**) is a snapshot of simulated results taken near the middle of the domain. The different colors in background in (**b**) indicate the wind speed contours. The white arrows represent wind vectors that indicate wind direction and magnitude of wind speed. The total height of the domain in the simulation is 50 m. The tree on lower-right corner of (**b**) illustrates the modeled canopy height. Note that the minimum wind speed is reached at the approximate height of the mid-canopy layer of maximum Leaf area index (LAI) and the secondary maximum wind speed is reached in the lower canopy trunk space

Assignment-2: Simulating 3-D Flow Past a Heated 3-D Cube on a Plate

Objectives:

This assignment will use a computational fluid dynamics (CFD) program to model the external cross flow over a 3-D object.

Key learning point:

- External simulation with appropriate domain sizes
- Buoyancy effect
- Influence of grid resolution.

Fig. 2.13 Comparison of the simulated wind profiles from the CFD model (solid line), the analytical solution (line with horizontal dashes), and observations (filled circles)

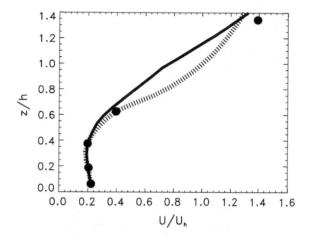

Fig. 2.14 Three CFD experiments based on the data collected at the Niwot Ridge Ameriflux site in the Rocky Mountains of Colorado under conditions: **a** cold inflow; **b** warm inflow; and **c** close to environmentally thermal condition

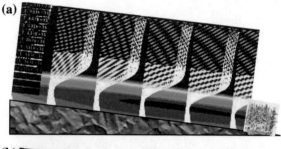

Simulation Steps:

(1) Build a cube (L = 1 unit) at the center of the flow domain;
(2) Select appropriate flow domain sizes to be modeled;
(3) Prescribe proper boundary conditions including inflow (given flow conditions U, T: ensure $Re = UL/\upsilon = 10^5$), outflow (given P_{atm} or assumed fully developed flow), ground (no-slip condition with standard wall function, adiabatic), sky and both sides (given P_{atm} or assumed symmetric);
(4) Select a turbulence model: the standard k-ε model;
(5) Select a thermal effect model: the Boussinesq approximation;
(6) Define convergence criterion: 0.1%;
(7) Set iteration: at least 1000 steps for steady simulation.

Cases to Be Simulated:

(1) Test three different grid resolutions (200, 400 and 800 K) for the adiabatic cube;
(2) Test the cube with total heat flux of 1000 Watts with the 800 K grid.

Report:

(1) Case descriptions: description of the cases
(2) Simulation details: computational domain, grid cells, convergence status

   • Figure of the grids used (on X-Y and X-Z planes);
   • Figure of simulation convergence records.

(3) Result and analysis

   • Figure of flow vectors at the XZ and XY middle planes of the cube;
   • Figure of pressure contours at the XZ and XY middle planes of the cube;
   • Figure of velocity contours at the XZ and XY middle planes of the cube;
   • Figure of temperature contours at the XZ and XY middle planes of the cube;
   • Evaluate the influences of grid resolution on simulation (e.g., pressure distributions at the surfaces);
   • Compare the flow and thermal characterizes between the adiabatic and heated cases (e.g., pressure distributions at surfaces, flow separation/vortex sizes around the body).

(4) Conclusions (findings, result implications, CFD experience and lessons, etc.)

References

Chen Q, Moser A (1991) Simulation of a multiple-nozzle diffuser. In: Proceedings of 12th AIVC conference, vol 2, pp 1–14
Durran DR (1990) Mountain waves and downslope winds. In: Blumen B (ed) Atmospheric processes over complex terrain. American Meteorological Society, Boston, pp 59–81
Hoener SF (1965) Fluid dynamics drag: practical information on aerodynamic drag and hydrodynamic resistance. Published by the Author, Midland Park, New Jersey

Launder BE, Spalding DB (1974) The numerical computation of turbulent flows. Comput Methods Appl Mech Eng 3:269–289

Raupach MR, Thom PG (1981) Turbulence in and above plant canopies. Ann Rev Fluid Mech 13:97–129

Siqueira M, Katul GG (2002) Estimating heat sources and fluxes in thermally stratified canopy flows using higher-order closure models. Bound Layer Meteorol 103:125–142

Turnipseed AA, Anderson DE, Burns S, Blanken PD, Monson RK (2004) Airflows and turbulent flux measurements in mountainous terrain, Part 2. Mesoscale effects. Agric Forest Meteorol 125:187–205

Yakhot V, Orzag SA, Thangam S, Gatski TB, Speziak CG (1992) Development of turbulence models for shear flows by a double expansion technique. Phys Fluids A 4:1510–1520

Yang C, Demokritou P, Chen Q, Spengler JD, Parsons A (2000) Ventilation and air quality in indoor ice skating arenas. ASHRAE Trans 106(2)

Yi C, Monson RK, Zhai Z, Anderson DE, Lamb B, Allwine G, Turnipseed AA, Burns SP (2005) Modeling and measuring the nocturnal drainage flow in a high-elevation, subalpine forest with complex terrain. J Geogr Res 110:D22303. https://doi.org/10.1029/2005JD006282

Zhai Z, Hamilton SD, Huang JM et al (2000) Integration of indoor and outdoor airflow study for natural ventilation design using CFD. In: Proceedings of the 21st AIVC annual conference on innovations in ventilation technology, The Hague, Netherlands

Zhai Z, Srebric J, Chen Q (2003) Application of CFD to predict and control chemical and biological agent dispersion in buildings. Int J Vent 2(3):251–264

Chapter 3
Select Equations to Be Solved

3.1 Fluid Mechanics Analysis System: Reynolds Transport Theorem

Fluid mechanics mainly applies three conservation laws of mass, energy, and momentum to the fluid flow (called governing equations of flow) and predicts the flow characteristics and the interactions among fluids as well as with solids. The mass, momentum, and energy conservation laws can be applied using two different analysis systems—closed system and open system.

(1) **Closed system (fixed mass or control mass system): Lagrangian System**

In a closed system, objects with fixed mass (e.g., a solid ball) are isolated and their changes in energy and momentum are tracked along with relevant properties such as pressure, velocity, temperature, etc (Fig. 3.1). The size and shape of the system may change during a process but there is no mass transfer in or out through the boundaries of this control mass. Closed systems are mostly used in thermodynamics and solid mechanics, where the state and movement of a certain object are the focus. For fluid mechanics, most cases are interested in the flow characteristics contained in a confined space rather than the pre- and post-fates of fluid before entering and after leaving the container. For instance, in indoor environment quality study, one may concentrate on the temperature and velocity distributions in a room caused by supply and exhaust diffusers, while ignoring where the air comes from and exhausts to. On the other hand, tracking the boundaries and movement of a fluid mass is much more challenging than tracking a solid mass due to irregularity and sometimes discontinuity of fluid geometry (such as when splattering). Nevertheless, a closed system may be used for some fluid flow problems such as tracking the trajectories of virus transportation in a space, where the specific objects (virus in this case) are the focused interest of the study.

(2) **Open system (fixed volume or control volume system): Euler System**

In an open system, the volume of a space is isolated and studied for the changes of mass, energy, and momentum in this volume during a process. The system allows the

© Springer Nature Singapore Pte Ltd. 2020
Z. Zhai, *Computational Fluid Dynamics for Built and Natural Environments*, https://doi.org/10.1007/978-981-32-9820-0_3

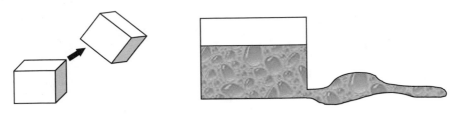

Fig. 3.1 Illustrations of using closed system for solid (left) and fluid (right)

flow in and out of mass, energy and momentum across the boundaries (called control surfaces) of the control volume. A control volume may also move and deform during a process, although most real-world applications utilize fixed and non-deformable control volumes to simplify the problems in study.

Another analogy to distinguish the control mass from the control volume system is to monitor students in a classroom. If the goal is to track the movement and properties of every student during the whole process (even before and after the class), one needs to track individual students (fixed mass) before they enter or after they leave the classroom. This is a closed system analysis. If the goal is to simply count student number in the classroom (without interest in where they come from and leave for), the classroom is the fixed or control volume to be explored and thus an open system should be used to allow students enter and leave the room.

(3) **Conversion from closed system to open system with Reynolds Transport Theorem**

Although different systems can be used to analyze the status of objects, the physics in conservation laws is uniform and independent of the analysis system selected. The Reynolds transport theorem (RTT) provides the link between the control mass and the control volume approaches, converting the conservation equations in one system to the other. The following provides the general format of the RTT:

$$\frac{dB_{CM}}{dt} = \frac{dB_{CV}}{dt} - \dot{B}_{in} + \dot{B}_{out} \qquad (3.1)$$

where B can be any variable (such as mass, energy, and momentum). CM stands for control mass and CV stands for control volume. $\dot{B}$ is the flow rate of variable in and out of the control volume.

3.2 Fluid Mechanics Conservation Equations in Integral Form

(1) **Mass Conservation**

If $B = M$ (mass), Eq. (3.1) becomes

$$\frac{dM_{CM}}{dt} = \frac{dM_{CV}}{dt} - \dot{M}_{in} + \dot{M}_{out} \qquad (3.2)$$

Since for a control mass, M does not change with time,

$$\frac{dM_{CM}}{dt} = 0 = \frac{dM_{CV}}{dt} - \dot{M}_{in} + \dot{M}_{out} \qquad (3.3)$$

The mass conservation equation for a control volume is thus

$$\frac{dM_{CV}}{dt} = \dot{M}_{in} - \dot{M}_{out} \qquad (3.4)$$

The mass change in a control volume is attributed to the imbalance between the mass flow in and the mass flow out. For a steady flow, $d/dt = 0$ for any variable,

$$\dot{M}_{in} = \dot{M}_{out} \qquad (3.5)$$

This $\dot{M}$ may imply the sum of multiple inlets and outlets to the control volume.

(2) **Energy Conservation**

If $B = E$ (total energy), Eq. (3.1) becomes

$$\frac{dE_{CM}}{dt} = \frac{dE_{CV}}{dt} - \dot{E}_{in} + \dot{E}_{out} \qquad (3.6)$$

where E is the total energy (including both mechanical and thermal energy); $E = H + E_P + E_K = U + PV + E_P + E_K$. $H = U + PV = mC_P T$ is enthalpy; $U = mC_V T$ is internal energy. C_P and C_V are specific heat at constant pressure and constant volume, respectively. T is temperature, P is pressure and V is volume. $E_P = mgz$ is potential energy, and $E_k = mv^2/2$ is kinetic energy (v is velocity). $\dot{E}$ is the flow rate of total energy in and out of the control volume.

The first law of thermodynamics states

$$\frac{dE_{CM}}{dt} = \dot{Q} + \dot{W} + \dot{S} \qquad (3.7)$$

where $\dot{Q}$ is the heat transfer (rate) imposed on the volume, $\dot{W}$ is the mechanical work (rate) conducted on the volume, and $\dot{S}$ is any additional energy (rate) occurred during the process (such as from chemical or nuclear reactions etc.). Hence, for a control volume (Fig. 3.2),

Fig. 3.2 Illustration of
energy conservation in a
control volume

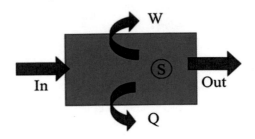

$$\frac{dE_{CV}}{dt} = \dot{E}_{in} - \dot{E}_{out} + \dot{Q} + \dot{W} + \dot{S} \tag{3.8}$$

(3) Momentum Conservation

If $B = M\vec{V}$(momentum), Eq. (3.1) provides

$$\frac{d(M\vec{V})_{CM}}{dt} = \frac{d(M\vec{V})_{CV}}{dt} - (\dot{M}\vec{V})_{in} + (\dot{M}\vec{V})_{out} \tag{3.9}$$

where $\vec{V}$ is the velocity vector, and $\dot{M}$ is the mass flow rate at inlet and outlet.
   The Newton's second law for a control mass states:

$$\frac{d(M\vec{V})_{CM}}{dt} = \sum \vec{F} \tag{3.10}$$

where the force $\vec{F}$ includes various body and surface forces such as gravity (body),
pressure (normal stress at surfaces), and friction (shear stress at surfaces) forces.
Combining Eqs. (3.9) and (3.10) yields

$$\frac{d(M\vec{V})_{CV}}{dt} - (\dot{M}\vec{V})_{in} + (\dot{M}\vec{V})_{out} = \sum \vec{F} \tag{3.11}$$

For a steady flow,

$$(\dot{M}\vec{V})_{out} - (\dot{M}\vec{V})_{in} = \sum \vec{F} \tag{3.12}$$

   Again, $\dot{M}\vec{V}$ here implies the summary of momentum forces at various inlets and
outlets of the control volume in study.

3.3 Fluid Mechanics Conservation Equations in Differential Form

The section above presents the fundamental flow governing equations in the integral form, which clearly reveals the principles of the conservation of mass, energy and momentum in fluid flow in a control volume. The integral expression of the governing equation is good for manual calculation for simplified flow problems, such as with steady, one-dimensional assumptions, and for computing average flow properties (e.g., one single temperature and velocity for the entire volume). To predict complex fluid flows with adequate spatial and temporal resolutions using a computer, the differential form of flow governing equations must be introduced and used.

To ease the writing and reading of a lengthy mathematic equation, the Einstein notation is often used in mathematics. The Einstein notation or Einstein summation convention is a notational convention that implies summation over a set of indexed terms in a formula, thus achieving notational brevity. It was introduced by Albert Einstein in 1916. According to this convention, when an index variable appears twice in a single term it implies summation of that term over all the values of the index (e.g., 1, 2, and 3 for a 3-D problem while 1 and 2 for a 2-D problem). Below are a few examples that are commonly seen in fluid mechanics:

- $U_i U_i = U_1 U_1 + U_2 U_2 + U_3 U_3$ ($U_1 = U$, $U_2 = V$, $U_3 = W$ in a 3-D flow)
- $\frac{dU_i}{dx_i} = \frac{dU_1}{dx_1} + \frac{dU_2}{dx_2} + \frac{dU_3}{dx_3}$ ($x_1 = x$, $x_2 = y$, $x_3 = z$ in a 3-D flow)
- $U_j \frac{dU_i}{dx_j} = U_1 \frac{dU_i}{dx_1} + U_2 \frac{dU_i}{dx_2} + U_3 \frac{dU_i}{dx_3}$ (where i can be any one but only one of 1, 2, 3)
- $\frac{\partial^2 \tau_{ij}}{\partial x_i \partial x_j} = \frac{\partial^2 \tau_{11}}{\partial x_1 \partial x_1} + \frac{\partial^2 \tau_{12}}{\partial x_1 \partial x_2} + \frac{\partial^2 \tau_{13}}{\partial x_1 \partial x_3} + \frac{\partial^2 \tau_{21}}{\partial x_2 \partial x_1} + \frac{\partial^2 \tau_{22}}{\partial x_2 \partial x_2} + \frac{\partial^2 \tau_{23}}{\partial x_2 \partial x_3} + \frac{\partial^2 \tau_{31}}{\partial x_3 \partial x_1} + \frac{\partial^2 \tau_{32}}{\partial x_3 \partial x_2} + \frac{\partial^2 \tau_{33}}{\partial x_3 \partial x_3}$
  (Here one pair of i and one pair of j appear and thus each of i and j should expand over all the values of the index).

For a single-phase Newtonian fluid (where viscosity does not depend on flow velocity and stress state), the general governing equations of flow may be expressed as below, in a Cartesian coordinate system.

(1) Continuity Equation (Mass Conservation)

$$\frac{\partial \rho}{\partial t} + \frac{\partial}{\partial x_j}(\rho u_j) = 0 \tag{3.13}$$

where, ρ is the air density, u_j is the instantaneous velocity component in three perpendicular coordinate directions (x_j, $j = 1, 2, 3$), and t is the time.

The following presents the derivation of the continuity equation. For an infinite small volume (or cell/mesh) dv, the cell center holds the velocity u, v, w at the coordinate of (x, y, z). The mass flow rate at the surface of $(x - 0.5dx)$ (called west surface) into the cell is thus:

Fig. 3.3 Illustration of mass
conservation over a control
volume dv = dxdydz

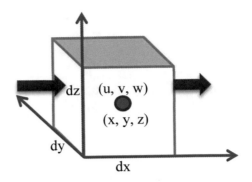

$$\dot{m}_{x-0.5dx} = \left[\rho u - \frac{\partial}{\partial x}(\rho u) \cdot \frac{1}{2}dx \right] \cdot dydz \tag{3.14}$$

And the mass flow rate at the surface of $(x + 0.5dx)$ (called east surface) out of
the cell is:

$$\dot{m}_{x+0.5dx} = \left[\rho u + \frac{\partial}{\partial x}(\rho u) \cdot \frac{1}{2}dx \right] \cdot dydz \tag{3.15}$$

dydz is the area of west and east surfaces as illustrated in Fig. 3.3. The net mass flow
rate on the X coordinate is then:

$$\dot{m}_{x-0.5dx} - \dot{m}_{x+0.5dx}$$

$$= \left[\rho u - \frac{\partial}{\partial x}(\rho u) \cdot \frac{1}{2}dx \right] \cdot dydz - \left[\rho u + \frac{\partial}{\partial x}(\rho u) \cdot \frac{1}{2}dx \right] \cdot dydz$$

$$= -\frac{\partial}{\partial x}(\rho u) \cdot dxdydz \tag{3.16}$$

Similarly, the net mass flow rate on the Y and Z coordinates can be obtained,
respectively:

$$\dot{m}_{y-0.5dy} - \dot{m}_{y+0.5dy}$$

$$= \left[\rho v - \frac{\partial}{\partial y}(\rho v) \cdot \frac{1}{2}dy \right] \cdot dxdz - \left[\rho v + \frac{\partial}{\partial y}(\rho v) \cdot \frac{1}{2}dy \right] \cdot dxdz$$

$$= -\frac{\partial}{\partial y}(\rho v) \cdot dxdydz \tag{3.17}$$

$$\dot{m}_{z-0.5dz} - \dot{m}_{z+0.5dz}$$

$$= \left[\rho w - \frac{\partial}{\partial z}(\rho w) \cdot \frac{1}{2}dz \right] \cdot dxdy - \left[\rho w + \frac{\partial}{\partial z}(\rho w) \cdot \frac{1}{2}dz \right] \cdot dxdy$$

$$= -\frac{\partial}{\partial z}(\rho w) \cdot dxdydz \tag{3.18}$$

The total mass change in the control volume $dv = dxdydz$ over the time is thus equal to:

$$\frac{\partial \rho}{\partial t}dxdydz = -\frac{\partial(\rho u)}{\partial x}dxdydz - \frac{\partial(\rho v)}{\partial y}dxdydz - \frac{\partial(\rho w)}{\partial z}dxdydz \tag{3.19}$$

$$\frac{\partial \rho}{\partial t} = -\frac{\partial(\rho u)}{\partial x} - \frac{\partial(\rho v)}{\partial y} - \frac{\partial(\rho w)}{\partial z} \tag{3.20}$$

Equation (3.20) is the same as Eq. (3.13), a general expression of the mass conservation of fluid flow.

For steady flows, Eq. (3.13) becomes:

$$\frac{\partial\left(\rho u_j\right)}{\partial x_j} = 0 \tag{3.21}$$

If considering incompressible fluids [i.e., the fluid density in the volume does not change during a flow process; however, the density may still be a function of space (x, y, z)], Eq. (3.13) can be expressed as:

$$\frac{\partial\left(\rho u_j\right)}{\partial x_j} = 0 \tag{3.22}$$

Note that Eqs. (3.21) and (3.22) are exactly the same; however, Eq. (3.22) does not imply a steady state flow, i.e., other variables such as velocity and temperature may still be able to vary with time. If assuming a constant fluid density, Eq. (3.22) can be further simplified as:

$$\frac{\partial u_j}{\partial x_j} = 0 \tag{3.23}$$

(2) Momentum Equations (Momentum Conservation)

$$\frac{\partial}{\partial t}(\rho u_i) + \frac{\partial}{\partial x_j}\left(\rho u_j u_i\right) = -\frac{\partial p}{\partial x_i} + \frac{\partial t_{ji}}{\partial x_j} + \rho F_i \tag{3.24}$$

where, u_i and u_j are, respectively, the instantaneous velocity component in x_i and x_j direction; p is the instantaneous pressure; t_{ij} is the component of viscous stress tensor; and F_i is the volume force working on the fluid.

The equation can be better understood in physics if a control volume $dv = dx_1 dx_2 dx_3$ is multiplied to each term in Eq. (3.24). Using $i = 1$ as a demonstration, the first term on the left is called *dynamic term*, which represents the change of the momentum ($dm \times u_1 = \rho dv \times u_1$) over the time in the control volume in the x_1 (or X) direction,

$$\frac{\partial}{\partial t}(\rho u_1)dv = \frac{\partial}{\partial t}(\rho dv \cdot u_1) \tag{3.25}$$

The second term on the left can be rewritten as:

$$\frac{\partial}{\partial x_j}(\rho u_j u_1)dv = \frac{\partial(\rho u_1 u_1)}{\partial x_1}dx_1 dx_2 dx_3 + \frac{\partial(\rho u_2 u_1)}{\partial x_2}dx_1 dx_2 dx_3$$

$$+ \frac{\partial(\rho u_3 u_1)}{\partial x_3}dx_1 dx_2 dx_3$$

$$= \partial[(\rho u_1 \cdot dx_2 dx_3) \cdot u_1] + \partial[(\rho u_2 \cdot dx_1 dx_3) \cdot u_1]$$

$$+ \partial[(\rho u_3 \cdot dx_1 dx_2) \cdot u_1]$$

$$= \partial(\dot{m}_1 \cdot u_1) + \partial(\dot{m}_2 \cdot u_1) + \partial(\dot{m}_3 \cdot u_1) \tag{3.26}$$

This represents the differences of the momentum entering and leaving the control volume, respectively, through the west ($x - 0.5dx$) and east ($x + 0.5dx$) surfaces, the south ($y - 0.5dy$) and north ($y + 0.5dy$) surfaces, and the bottom ($z - 0.5dz$) and top ($z + 0.5dz$) surface. $\dot{m}_1$, $\dot{m}_2$, $\dot{m}_3$ are the actual mass flow rates entering and leaving the cell (calculated using the velocity normal to the surfaces) at x, y, and z· directions. Each of these mass flow rates may bring the momentum impacts to the control volume on the x_1 (or X) direction via the x_1 direction velocity component u_1 at each surface. The same physics is shown in the integral Eq. (3.11). This term is called *convection or advection term* as it is directly related to fluid flow.

The first term on the right of Eq. (3.24) is named *pressure term*, which represents the pressure forces acted on the cell surfaces that drive the flow. Equation (3.27) shows the pressure forces on the west and east surfaces that affect the momentum $\rho dv \cdot u_1$ in x_1 direction,

$$-\frac{\partial p}{\partial x_1}dv = -\frac{\partial p}{\partial x_1}dx_1 dx_2 dx_3 = -\partial(p dx_2 dx_3) \tag{3.27}$$

The second term on the right of Eq. (3.24) is the impact from viscous stresses/forces at the surfaces of the volume and t_{ij} is viscous stress tensor.

$$\frac{\partial t_{j1}}{\partial x_j}dv = \partial(t_{11}dx_2 dx_3) + \partial(t_{21}dx_1 dx_3) + \partial(t_{31}dx_1 dx_2) \tag{3.28}$$

where the first term on the right is the normal stress influence and the other two are the shear stress influence on the momentum $\rho dv \cdot u_1$ in x_1 direction.

According to the Stokes' law, the viscous stress tensor t_{ij} can be represented as:

$$t_{ij} = \mu \left(\frac{\partial u_i}{\partial x_j} + \frac{\partial u_j}{\partial x_i} \right) - \frac{2}{3} \mu \frac{\partial u_k}{\partial x_k} \delta_{ij} \tag{3.29}$$

where μ is the molecular dynamic viscosity. $\delta_{ij} = 1$ if $i = j$ (otherwise zero). Note that the relationship of Eq. (3.29) only works for Newtonian fluids (e.g., air and water). The general expression of the momentum Eq. (3.24) is applicable for all fluids but may have different stress-strain correlations that are attributed to inherent properties of fluids.

The last term on Eq. (3.24) represents the body force on each volume/cell, which can be gravity or magnetic force etc. If gravity is considered, this *source term* can be written as

$$\rho F_i dv = \rho dv \cdot g_i \tag{3.30}$$

where $F_i = g_i$ is the gravitational acceleration in the x_i direction.

(3) **Energy Equations (Energy Conservation)**

$$\frac{\partial}{\partial t} \left[\rho \left(e + \frac{u_i u_i}{2} \right) \right] + \frac{\partial}{\partial x_j} \left[\rho u_j \left(e + \frac{u_i u_i}{2} \right) \right] = \frac{\partial}{\partial x_j} (u_i t_{ij})$$

$$- \frac{\partial}{\partial x_j} (p u_j) + \rho F_i u_i - \frac{\partial q_i}{\partial x_i} + \rho q_{source} \tag{3.31}$$

where e is the internal energy of the fluid (unit: kJ/kg), $\frac{u_i u_i}{2}$ is the instantaneous kinetic energy of the fluid, q_i is the heat flux in in x_i direction, and q_{source} is the energy source in the fluid.

The first term on the left of the equation is the dynamic term and is the total energy change within the control volume over the time. The second term is the convection term, representing the energy with flows entering/leaving the volume through the surfaces. The first term on the right is the energy from the mechanical work caused by surface stresses (e.g., frictions); the second term is the energy from the mechanical work by pressure (e.g., either pressure changes at the cell surfaces or cell volume change); the third one is the energy from the mechanical work done by the body force (e.g., gravity); the fourth one is the heat transfer across the surfaces of the volume; and the last one represents other energy sources in the volume (e.g., from chemical reactions inside the volume).

If using the enthalpy h to replace the internal energy e (i.e., the PV work is considered in the fluid total energy, which is common), Eq. (3.31) becomes

$$\frac{\partial}{\partial t} \left[\rho \left(h + \frac{u_i u_i}{2} \right) \right] + \frac{\partial}{\partial x_j} \left[\rho u_j \left(h + \frac{u_i u_i}{2} \right) \right] = \frac{\partial}{\partial x_j} (u_i t_{ij}) + \rho F_i u_i$$

$$- \frac{\partial q_i}{\partial x_i} + \rho q_{source} + \frac{\partial p}{\partial t} \tag{3.32}$$

Typically, the following terms are grouped as a source term for the energy equation,

$$\phi = \frac{\partial}{\partial x_j}\left(u_i t_{ij}\right) + \rho F_i u_i + \rho q_{source} + \frac{\partial p}{\partial t} \tag{3.33}$$

The right term 1, 2 and 4 are generally smaller than the heat transfer term $-\frac{\partial q_i}{\partial x_i}$ and thus neglected by many practical CFD software and simulations. Hence, $\phi = \rho q_{source}$. The change of instantaneous kinetic energy $\frac{u_i u_i}{2}$ is also smaller compared to either internal energy or enthalpy, and hence often ignored in the energy equation. The refined energy equation then becomes:

$$\frac{\partial(\rho h)}{\partial t} + \frac{\partial\left(\rho u_j h\right)}{\partial x_j} = -\frac{\partial q_i}{\partial x_i} + \phi \tag{3.34}$$

For ideal gases and incompressible fluids, the enthalpy of fluid can be calculated by:

$$h = C_p T \tag{3.35}$$

where C_p is specific heat at constant pressure (and usually treated as a constant), and T is the instantaneous fluid temperature. According to the Fourier's law, the conductive heat transfer in the fluid can be expressed as:

$$q_i = -\kappa \frac{\partial T}{\partial x_i} \tag{3.36}$$

where k is the thermal conductivity of fluid. Substituting (3.35) and (3.36) into (3.34) yields

$$\frac{\partial\left(\rho C_p T\right)}{\partial t} + \frac{\partial\left(\rho C_p u_j T\right)}{\partial x_j} = \frac{\partial}{\partial x_k}\left(\kappa \frac{\partial T}{\partial x_k}\right) + \phi \tag{3.37}$$

The flow governing Eqs. (3.13), (3.24) and (3.31) are generally called the Navier-Stokes equations. Equations (3.13), (3.24) and (3.37) forms a complete set of flow governing equations for ideal gases and incompressible fluids—two commonly encountered flows in fluid engineering applications, with five (5) equations for six (6) variables: u_1, u_2, u_3, p, T, ρ. Additional equation is required to enclose this problem mathematically. For ideal gases, it is the state equation,

$$p = \rho R T \tag{3.38}$$

where R is the ideal gas constant.

(a) *Instantaneous Governing Equations for Ideal Gas Flows*

$$\frac{\partial \rho}{\partial t} + \frac{\partial (\rho u_j)}{\partial x_j} = 0 \qquad (3.39)$$

$$\frac{\partial (\rho u_i)}{\partial t} + \frac{\partial (\rho u_j u_i)}{\partial x_j} = -\frac{\partial p}{\partial x_i} + \frac{\partial t_{ji}}{\partial x_j} + \rho g_i \qquad (3.40)$$

$$\frac{\partial (\rho C_p T)}{\partial t} + \frac{\partial (\rho C_p u_j T)}{\partial x_j} = \frac{\partial}{\partial x_k}\left(\kappa \frac{\partial T}{\partial x_k}\right) + \phi \qquad (3.41)$$

$$p = \rho RT \qquad (3.42)$$

(b) *Instantaneous Governing Equations for Incompressible Fluid Flows*

Equations (3.13), (3.24) and (3.37) can also be closed by using the incompressible assumption for fluids, where the fluid density ρ is assumed to be constant.

By substituting the t_{ij} expression (3.29) into Eq. (3.24) and taking into account the continuity Eq. (3.23) for incompressible fluids and assuming gravity is the only body force, the momentum Eq. (3.24) can be rewritten as

$$
\begin{aligned}
\frac{\partial}{\partial t}(\rho u_i) + \frac{\partial}{\partial x_j}(\rho u_j u_i) &= -\frac{\partial p}{\partial x_i} + \frac{\partial}{\partial x_j}\left(\mu\left(\frac{\partial u_i}{\partial x_j} + \frac{\partial u_j}{\partial x_i}\right)\right) + \rho g_i \\
&= -\frac{\partial p}{\partial x_i} + \mu\frac{\partial}{\partial x_j}\left(\frac{\partial u_i}{\partial x_j}\right) + \mu\frac{\partial}{\partial x_j}\left(\frac{\partial u_j}{\partial x_i}\right) + \rho g_i \\
&= -\frac{\partial p}{\partial x_i} + \mu\frac{\partial}{\partial x_j}\left(\frac{\partial u_i}{\partial x_j}\right) + \mu\frac{\partial}{\partial x_i}\left(\frac{\partial u_j}{\partial x_j}\right) + \rho g_i \\
&= -\frac{\partial p}{\partial x_i} + \mu\frac{\partial}{\partial x_j}\left(\frac{\partial u_i}{\partial x_j}\right) + \rho g_i
\end{aligned} \qquad (3.43)
$$

Since the fluid density is treated as constant, the influence of fluid temperature variation on the density and then on the flow momentum, in terms of buoyancy, is decoupled. As a result, the Boussinesq buoyancy approximation is suggested to couple the momentum and energy equations. As a first order truncation, the Boussinesq buoyancy approximation presents the relationship between gas density and temperature as

$$\rho = \rho_0[1 - \beta(T - T_o)] \qquad (3.44)$$

where ρ_o is the reference density at the reference temperature T_o. $\beta = 1/T$ is the coefficient of volume expansion of the fluid (unit: 1/K). Taking this into Eq. (3.43) and absorbing the constant $\rho_0 g_i$ into the pressure term (because only the pressure difference matters), the momentum equation for incompressible fluids becomes

$$\frac{\partial}{\partial t}(u_i) + \frac{\partial}{\partial x_j}(u_j u_i) = \frac{\partial}{\partial t}(u_i) + u_j \frac{\partial}{\partial x_j}(u_i)$$

$$= -\frac{\partial p}{\rho \partial x_i} + \frac{\mu}{\rho} \frac{\partial}{\partial x_j}\left(\frac{\partial u_i}{\partial x_j}\right) - g_i \beta (T - T_o)$$

$$= -\frac{\partial p}{\rho \partial x_i} + \nu \frac{\partial^2 u_i}{\partial x_j \partial x_j} - g_i \beta (T - T_o) \qquad (3.45)$$

where $\nu = \mu / \rho$ is the kinematic viscosity (unit: m^2/s).

Energy Eq. (3.37) can also be revised as

$$\frac{\partial T}{\partial t} + \frac{\partial (u_j T)}{\partial x_j} = \frac{\partial T}{\partial t} + u_j \frac{\partial T}{\partial x_j} = \frac{1}{\rho c_p} \frac{\partial}{\partial x_k}\left(\kappa \frac{\partial T}{\partial x_k}\right) + \frac{\phi}{\rho c_p} \qquad (3.46)$$

The following is the complete set of governing equations for incompressible fluid flows with five (5) equations for five (5) variables: u_1, u_2, u_3, p, T.

$$\frac{\partial u_i}{\partial x_j} = 0 \qquad (3.47)$$

$$\frac{\partial u_i}{\partial t} + u_j \frac{\partial u_i}{\partial x_j} = -\frac{\partial p}{\rho \partial x_i} + \nu \frac{\partial^2 u_i}{\partial x_j \partial x_j} - g_i \beta (T - T_o) \qquad (3.48)$$

$$\frac{\partial T}{\partial t} + u_j \frac{\partial T}{\partial x_j} = \frac{1}{\rho c_p} \frac{\partial}{\partial x_k}\left(\kappa \frac{\partial T}{\partial x_k}\right) + \frac{\phi}{\rho c_p} \qquad (3.49)$$

(4)   General Scalar Transport Equation (Mass Conservation)

If a concentration of particular species (other than the domain fluid) is concerned in the flow, such as the concentrations of moisture and pollutant in the air, the concentration transport equation need be resolved. The concentration equation fundamentally is a mass transport or conservation equation, which can be expressed in a general scalar transport equation form as below:

$$\frac{\partial (\rho C)}{\partial t} + \frac{\partial (\rho u_j C)}{\partial x_j} = \frac{\partial}{\partial x_k}\left(\alpha \frac{\partial (C)}{\partial x_k}\right) + q_{source} \qquad (3.50)$$

where, C is the instantaneous scalar variable such as species concentration, α is the molecular diffusion coefficient for the scalar, and q_{source} is the source term. Note that Eq. (3.50) is very similar to the energy Eq. (3.37). Using contaminant concentration as an example, if the unit of C is kg_c/kg_{air}, the integration of the first term over the volume dv provides

$$\frac{\partial}{\partial t}(\rho C)dv = \frac{\partial}{\partial t}(\rho dv \cdot C) \quad \left(\text{unit: } kg_c/s\right) \qquad (3.51)$$

This is the change rate of the contaminant mass in dv. This change is due to

(1) the contaminant mass entering and leaving dv with the flow:

$$\frac{\partial}{\partial x_j}(\rho u_j C)dv = \frac{\partial(\rho u_1 C)}{\partial x_1}dx_1 dx_2 dx_3 + \frac{\partial(\rho u_2 C)}{\partial x_2}dx_1 dx_2 dx_3$$

$$+ \frac{\partial(\rho u_3 C)}{\partial x_3}dx_1 dx_2 dx_3$$

$$= \partial[(\rho u_1 \cdot dx_2 dx_3) \cdot C] + \partial[(\rho u_2 \cdot dx_1 dx_3) \cdot C]$$

$$+ \partial[(\rho u_3 \cdot dx_1 dx_2) \cdot C]$$

$$= \partial(\dot{m}_1 \cdot C) + \partial(\dot{m}_2 \cdot C) + \partial(\dot{m}_3 \cdot C) \quad (\text{unit: } kg_c/s) \quad (3.52)$$

(2) the dispersion (or diffusion) at the volume surfaces due to the concentration gradient:

$$\frac{\partial}{\partial x_k}\left(\alpha \frac{\partial C}{\partial x_k}\right)dv = \frac{\partial}{\partial x_1}\left(\alpha \frac{\partial C}{\partial x_1}\right)dv + \frac{\partial}{\partial x_2}\left(\alpha \frac{\partial C}{\partial x_2}\right)dv + \frac{\partial}{\partial x_3}\left(\alpha \frac{\partial C}{\partial x_3}\right)dv$$

$$= \left(\alpha \frac{\partial C}{\partial x_1}dA\right)_{east} - \left(\alpha \frac{\partial C}{\partial x_1}dA\right)_{west}$$

$$+ \left(\alpha \frac{\partial C}{\partial x_2}dA\right)_{north} - \left(\alpha \frac{\partial C}{\partial x_2}dA\right)_{south}$$

$$+ \left(\alpha \frac{\partial C}{\partial x_3}dA\right)_{top} - \left(\alpha \frac{\partial C}{\partial x_3}dA\right)_{bottom} \quad (3.53)$$

The molecular diffusion coefficient α has the same unit as dynamic viscosity μ, kg/(m s). $\alpha \frac{\partial C}{\partial x_i}dA$ is the diffusion at the surfaces due to the concentration gradient, and the unit of this is

$$\frac{kg_{air}}{m \cdot s} \cdot \frac{kg_c}{kg_{air}} \cdot \frac{1}{m} \cdot m^2 = \frac{kg_c}{s} \quad (3.54)$$

(3) the source term $q_{source} \cdot dv$ in unit of kg_c/s (i.e., either source or sink of the contaminant within the volume dv. Note that the scalar unit will vary according to the unit of the source term.

(5) **Uniform Expression of Flow Governing Equations**

The governing equations of incompressible flow (3.47)–(3.49) and the scalar transport Eq. (3.50) can be generalized into the following form:

$$\frac{\partial(\rho\phi)}{\partial t} + \frac{\partial(\rho U_j \phi)}{\partial x_j} = \frac{\partial}{\partial x_j}\left(\Gamma_{\phi,eff}\frac{\partial \phi}{\partial x_j}\right) + S_\phi \quad (3.55)$$

Table 3.1 Formula for the general form Eq. (3.55)

Equation	ϕ	$\Gamma_{\phi,eff}$	S_ϕ
Continuity	1	0	0
Momentum	U_i	μ	$-\frac{\partial p}{\partial x_i} - \rho\beta(T - T_\infty)g_i$
Temperature	T	$\frac{\mu}{Pr}$	S_T
Concentration	C	$\frac{\mu}{Sc}$	S_C

where ϕ represents the physical variable in question, as shown in Table 3.1. The equation has dynamic, convection, diffusion and source terms.

When selecting proper equations to be computed, appropriate assumptions (e.g., steady state and/or impressible state) and case simplifications (e.g., 2-D and constant coefficients) should be determined first. This will identify how many variables and equations to solve. Additional equations for temperature and concentrations should be included whenever the physics of flow requires so. More equations selected will impose extra computing sophistication and costs. Boundary (and initial) conditions are mandatory for all these equations as will be described in Chap. 6.

3.4 Transport Equations for Particle and Droplet

Predicting particle and droplet transport behaviors in the air is essentially a simulation of air-particle two-phase flows with continuous gas phase of air and dispersed solid/liquid phase of particles/droplets. To simulate the movement of continuous air, the flow governing Eqs. (3.47)–(3.49) in Eulerian-form are solved. To predict the transport of dispersed particles and droplets in the air, three kinds of models are usually available (Liu and Zhai 2007):

- Lazy particle model
- Isothermal particle model
- Vaporizing droplet model.

The lazy particle model does not solve the particle trajectories directly and thus does not produce individual particle velocities. It simply follows the continuous-phase velocity (streamline) at each point of the flow field—a model that is tracer-like (hence also called the 'tracer' model). The model does not handle either size or temperature of particles, and cannot undergo any physical process (e.g., solidification and vaporization) except turbulent dispersion. Lazy particles will not affect the continuous-phase solution. The distributed concentration of lazy particles can be simulated by solving the same transport equation for gas-phase contaminants, i.e. Eq. (3.50). Lazy particle model may be appropriate for small particles with quasi-gaseous compounds that have similar molecular weights to the elements in air and when particle-particle interaction is not concerned. Due to its simplicity, the model has been broadly used for indoor and outdoor particle study.

The isothermal particle model predicts particle trajectories and velocities by solving additional Lagrangian transport equations for particles without considering particle thermal effect. As a result, simulated particles do not change their sizes during the transport and there is no exchange of heat and mass between continuous and dispersed phases. The model can be applicable to many solid particle pollutants, such as, tobacco smoke particulates, soot, and fibers.

When liquid particles (droplets) are simulated, the evaporation effect could be important so that the Lagrangian particle transport equations must consider the exchange of mass and heat (besides momentum) between dispersed and continuous phases. This vaporizing droplet model represents the true physics of droplet dispersion but also complicates the simulation. Evaporation is commonly included in fire extinguishing modeling when water sprinklers are used, but very few researchers take this into account in air quality study because usually less mass and heat transfer occur during regular room-temperature droplet transport process. However, this small mass and heat transfer could be significant, such as for predicting the fate of droplets carrying viruses or bacteria.

Most studies take the liberty of deciding appropriate (or convenient) simulators to predict particle and droplet transport behaviors. Generally, lazy particle model is employed if one thinks that the particle size is relatively small and the distributed contaminant concentration is the major concern; otherwise, isothermal particle model will be utilized. Vaporizing droplet model is the least used model due to the complexity of the model unless special considerations need be taken into account. It is unclear which model is the most effective and efficient for a certain particle or droplet or how large is the difference in the results predicted by different models. It is also uncertain under what circumstances the air-particle interaction and droplet mass change must be considered as they influence the motion of particles and droplets in the space. The answers to these questions are affected by many factors, such as, simulation accuracy requirement, computing cost affordability, particle sizes, and environment conditions, etc. The following sections attempt to provide some practical particle-size-based criteria for selecting an appropriate particle simulation method.

(1) Theoretical Analysis

• *General Particle Transport Equation*

When using the Lagrangian method, the trajectory of each particle in the air can be computed by solving the momentum equation based on Newton's second law,

$$\frac{d(m\vec{v})}{dt} = \sum \vec{F} \tag{3.56}$$

and

$$\frac{d\vec{X}_p}{dt} = \vec{v} \tag{3.57}$$

where $\vec{v}$ is the particle velocity, m is the particle mass, $\sum \vec{F}$ stands for the total forces acted on the particle, and $\vec{X}_p$ is the co-ordinates of the particle. Momentum is transferred between air and particles through inter-phase drag and lift forces that can be divided into, but not limited to, the following parts: the drag force, pressure gradient and buoyancy forces, unsteady forces that include Basset force and virtual mass force, Brownian force, and body forces such as gravity force (Crowe et al. 1998). For particles with a certain size and density, some of the forces may be very small compared to others, and thus can be ignored. For most spherical particles, the particle motion equation can be simplified as (Crowe et al. 1998):

$$\frac{d(m\vec{v})}{dt} = \frac{1}{2}C_D \frac{\pi D^2}{4} \rho_a(\vec{u}-\vec{v})|\vec{u}-\vec{v}| + m\vec{g} \tag{3.58}$$

where $\vec{v}$ is the particle velocity, $\vec{u}$ is the local air velocity, D is the particle diameter, ρ_a is the air density, and C_D is the drag coefficient. Equation (3.58) only includes the most important drag force and gravity force acted on a particle, which is appropriate for particles with size above 1 μm and density above the order of 1 0^3 kg/m^3 (Jiang 2002).

By introducing the particle Reynolds number $Re_r = \frac{\rho_a D|\vec{u}-\vec{v}|}{\mu_a}$ and the spherical particle mass m $= \frac{1}{6}\rho_p \pi D^3$, Eq. (3.58) can be written as

$$\frac{d\vec{v}}{dt} = \frac{18\mu_a}{\rho_p D^2} \frac{C_D Re_r}{24}(\vec{u}-\vec{v}) + \vec{g} = \frac{18\mu_a}{\rho_p D^2} f(\vec{u}-\vec{v}) + \vec{g} \tag{3.59}$$

where μ_a is the air viscosity and ρ_p is the particle density. $f = \frac{C_D Re_r}{24}$ is defined as drag factor. One accurate correlation for f over the entire sub-critical Reynolds number range was developed by Clift and Gauvin (1970):

$$f = 1 + 0.15Re_r^{0.687} + 0.0175 \times \left(1 + 4.25 \times 10^4 Re_r^{-1.16}\right)^{-1} \tag{3.60}$$

Figure 3.4 illustrates the relationship between f and Re_r in Eq. (3.60), which shows that f approaches to 1 as $Re_r < 1$.

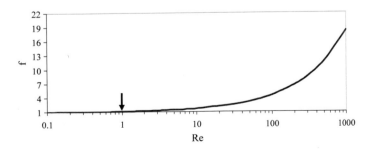

Fig. 3.4 Relationship between f and Re_r in Eq. (3.60)

• *Movement of Particle without Evaporation*

For spherical particles without considering mass change, particle diameters remain constant during the transmission. This analyze discusses the transport behaviors of such particles of different sizes in both low and high Reynolds flows.

(1) *Low Reynolds Flow ($Re_r < 1$)*

Since the drag factor f in Eq. (3.59) approximates to one when $Re_r < 1$, Eq. (3.59) can be expressed as:

$$\frac{d\vec{v}}{dt} = \frac{(\vec{u} - \vec{v})}{\tau_v} + \vec{g} \tag{3.61}$$

where $\tau_v = \frac{\rho_p D^2}{18\mu_a}$ is defined as the particle momentum (velocity) response time (second) and is constant for a certain particle with constant diameter.

To simplify the theoretical analysis, a two-dimensional air-particle flow is assumed in a x-y Cartesian coordinate system. Equation (3.61) then becomes

$$\begin{cases} \frac{dv_x}{dt} = \frac{u_x - v_x}{\tau_v} \\ \frac{dv_y}{dt} = \frac{u_y - v_y}{\tau_v} - g \end{cases} \tag{3.62}$$

By assuming a constant airflow velocity $\vec{u} = u_x\vec{i} + u_y\vec{j}$ and a zero initial particle velocity, the analytical solutions to Eq. (3.62) can be obtained

$$\begin{cases} v_x = u_x\left(1 - e^{-t/\tau_v}\right) \\ v_y = \left(u_y - \tau_v g\right)\left(1 - e^{-t/\tau_v}\right) \end{cases} \tag{3.63}$$

As a result, $v_x = u_x$ and $v_y = u_y - \tau_v g$ if $t = \infty$, and $v_x = 63\% u_x$ and $v_y = 63\%\left(u_y - \tau_v g\right)$ if $t = \tau_v$. Hence, the particle momentum response time τ_v indicates how fast the particle can reach the air velocity and respond to the air velocity changes. Figure 3.5 presents the change of particle momentum response time with particle diameters in the air. If τ_v is adequately small, particles can easily follow the air velocity so that lazy particle model is appropriate. Conversely, if τ_v is significantly large, the time needed to reach the air velocity is much longer than the time needed for particle to fall on floor and thus a free dropping calculation may be sufficient.

To find the critical particle momentum response times or particle diameters, the Stokes Number is introduced

$$St_v = \frac{\tau_v}{\tau_F} \tag{3.64}$$

where τ_F is the characteristic time of a flow field that represents the shortest time for a certain particle to be caught by obstructions. For indoor particles dispersing in a ventilated space as illustrated in Fig. 3.6, τ_F can be calculated via

Fig. 3.5 Relationship
between τ_v and particle
diameter D

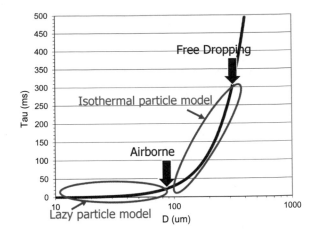

Fig. 3.6 Characteristic time
for a ventilated indoor space

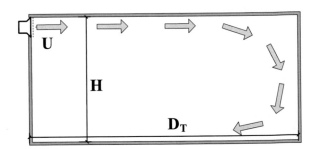

$$\tau_F = \min\left(D_T/U, \sqrt{2H/g}\right) \tag{3.65}$$

where D_T is the depth of the room, U is the vent inlet air velocity, H is the height of the room, g is the gravitational acceleration.

If the Stokes Number is far less than one, i.e., the particle momentum response time is much less than the characteristic time associated with the flow field, the particles will have ample time to respond to and follow the changes in flow velocity. As a result, the particle and fluid velocities can be treated as velocity equilibrium and lazy particle model can be applied. In contrast, if the Stokes Number is far larger than one, the particle will essentially have no time to respond to the fluid velocity changes before they are caught by building envelopes. For particles with St_v number in between, Lagrangian particle transport equation must be solved.

(2) *High Reynolds Flow ($Re_r > 1$)*

When $Re_r > 1$, $f = C_D Re_r/24$ is not equal to one any more. Equation (3.59) becomes

$$\frac{d\vec{v}}{dt} = \frac{(\vec{u} - \vec{v})}{\tau_v'} + \vec{g} \tag{3.66}$$

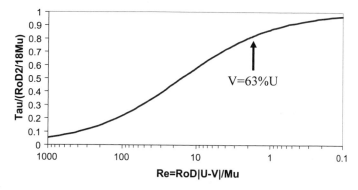

Fig. 3.7   Variation of $\tau_v'/(\rho_p D^2/18\mu_a) = 1/f$ with $Re_r = \rho_a D|\vec{u} - \vec{v}|/\mu_a$

where $\tau_v' = \frac{\rho_p D^2}{18\mu_a}\frac{1}{f}$ is defined as the modified particle momentum (velocity) response time. Because $f > 1$ for the entire subcritical Reynolds number range according to Eq. (3.60), τ_v' is always smaller than τ_v. Figure 3.7 illustrates the variation of $\tau_v'/(\rho_p D^2/18\mu_a) = 1/f$ with $Re_r = \rho_a D|\vec{u} - \vec{v}|/\mu_a$. When particle and air have relatively large velocity difference thus large Re_r, a large drag factor f occurs to change the particle velocity to follow the air speed, which corresponds to a small "local" modified particle momentum response time τ_v'. When the particle speed approaches the air velocity, less drag force is imposed on the particle, which leads to longer time to further alter its speed towards that of the free air. To be consistent with low Reynolds flow and produce a simple justification criterion, a "local" $\tau_{v=63\%U}'$ is used to represent the total time for a particle released from rest to achieving 63% of the free stream velocity. This number overestimates the real time but reflects its magnitude. By using the Stokes number $St_v = \tau_{v=63\%U}'/\tau_F$, the same conclusions as for low Reynolds flows can be reached for high Reynolds flows.

• *Movement of Particle with Evaporation*

For spherical particles with evaporation, their diameters keep varying due to evaporation during the transmission. Besides the particle momentum Eq. (3.59), an additional equation that describes such mass change of the droplet must be solved. One of the representative droplet mass change equations was developed by Ludwig et al. (2004):

$$\frac{dm_p}{dt} = -\pi D_p \frac{K_v}{Cp_v} Nu \ln(1 + B_m) \tag{3.67}$$

where m_p is the droplet mass, D_p is the droplet diameter, K_v is the thermal conductivity of droplet vapor, Cp_v is the specific heat capacity of droplet vapor. Nu is the Nusselt number, determined from the following correlation:

$$Nu = 2(1 + 0.3Re_r^{0.5} Pr^{0.33})F \tag{3.68}$$

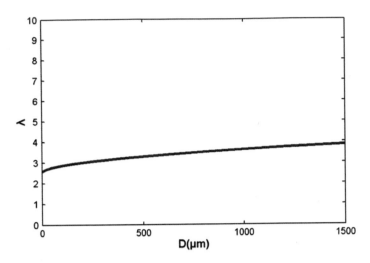

Fig. 3.8 Evaporation constant λ versus water droplet diameter

where Pr is the laminar Prandtl number of air and F is the Frossling correction for mass transfer given by $F = \ln(1 + B_m)/B_m$. B_M is the mass transfer number, which represents the "driving force" in the mass transfer process, and is defined by:

$$B_M = \left[\frac{Y_{vs} - Y_{v\infty}}{1 - Y_{vs}}\right] \qquad (3.69)$$

where $Y_{v\infty}$ is the mass fraction of droplet vapour in the air surrounding the droplet, and Y_{vs} is the mass fraction of droplet vapour at the surface of droplet and can be calculated via:

$$Y_{vs} = \left[1 + \left(\frac{P}{P_{vs}} - 1\right)\frac{W_a}{W_v}\right]^{-1} \qquad (3.70)$$

where P is the total pressure of air surrounding droplet, P_{vs} is the partial pressure of droplet vapour at the surface of droplet at the saturation conditions defined by the droplet temperature, W_a is the molecular weight of air, and W_v is the molecular weight of droplet vapour.

Equation (3.67) can be rewritten as

$$D_p\frac{dD_p}{dt} = -2\frac{K_v}{Cp_v}\frac{Nu}{\rho_p}\ln(1 + B_m) = -\frac{\lambda}{2} \qquad (3.71)$$

Numerical experiments show that $\lambda = 4\frac{K_v}{Cp_v}\frac{Nu}{\rho_p}\ln(1 + B_m)$ is almost constant for a certain droplet vapor under typical room conditions (and so called the evaporation constant). Figure 3.8 verifies that λ increases less than 1.7 times when water droplet diameter changes from nearly 0 to 1500 μm with a droplet temperature of 310 K.

Integrating Eq. (3.71) with the evaporation constant thus provides

$$D^2 = D_0^2 - \lambda t \tag{3.72}$$

This is a popular form of the evaporation equation that has been extensively used in the past. The evaporation lifetime of a droplet is then defined as the time needed to change droplet diameter from D_0 to $D = 0$

$$\tau_m = \frac{D_0^2}{\lambda} \tag{3.73}$$

To quantify the relative evaporation speed of a droplet, a new index—evaporation effectiveness (EE) number—has been introduced

$$EE = \frac{\tau_m}{\tau_F} = \frac{D_0^2}{\lambda \tau_F} \tag{3.74}$$

If $EE \ll 1$, i.e., the droplet evaporation time is much less than the characteristic time associated with the flow field, the droplet evaporates and disappears very fast. Therefore, such droplets can be treated as airborne. Conversely, if $EE \ll 1$, the droplet will be caught by building enclosures before it barely changes its diameter through evaporation. In this case, the evaporation-free particle model will be sufficient. For all other cases with EE numbers falling in between, a particle model with evaporation has to be considered.

(2) **Numerical Experiments**

• *Case Descriptions*

Figure 3.9 shows the two-dimensional ventilated room modeled with CFD. The room is 10 m long and 3 m high with supply inlet at the top left corner of the room and exhaust vent at the bottom right corner. The supply air velocity is 0.1 m/s. A still particle or droplet is released from the center of the room at height = 1.8 m (nose level). The flow characteristic time for this case is 0.6 s, which is the dropping time of a free object from 1.8 m.

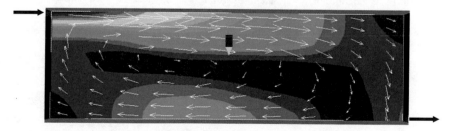

Fig. 3.9 Simulated flow pattern of the ventilated room

• *Particle without Evaporation*

Figure 3.10 shows the predicted trajectories of isothermal particles with different diameters by CFD. For comparison, the 1500 s trajectory of lazy particles released at the same location is illustrated, which represents the streamline of airflow through the source location. The 20 μm particle can be seen as airborne because it never falls onto the floor and tends to follow the airflow. The 40 μm particle will hit the floor after 37 s while the 100 μm particle will do so after 7.2 s. For both cases, the influence of airflow on particle trajectories is perceivable. The trajectories and falling time of the 1500 μm particle and 10,000 μm particle are almost identical. The 10,000 μm-particle is almost like an object in free-fall since the dropping time is very close to 0.6 s—the flow characteristic time. Therefore, for typical room conditions, 20 and 1500 μm can be used as critical diameters between airborne particles, Lagrangian isothermal particles and free dropping particles, which correspond to the Stokes numbers of 0.001 and 10, respectively, as demonstrated in Table 3.2.

• *Particle with Evaporation*

Figure 3.11 shows the predicted trajectories of vaporizing particles (droplets) with different diameters by CFD. In the simulation, the room air temperature remains 293 K, while the initial droplet temperature is the same as the normal human body temperature of 310 K. The results show that the 40 μm droplet completely evaporates to the air after 2.18 s before it starts to spread while the 100 μm droplet takes 12.2 s to fully evaporate. The 300 μm droplet falls on the floor after 2.31 s during which

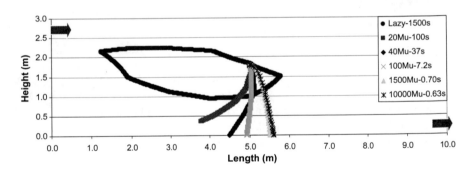

Fig. 3.10 Predicted trajectories of isothermal particles with different sizes in a ventilated room

Table 3.2 Model selection criteria for simulating isothermal particle movement in the air

Stokes number	Category	Critical St_v	Corresponding D
$St_v \ll 1$	Lazy particle	$St_{v,cr} = 0.001$	$D_{cr} = 20\,\mu m$
$St_v \approx 1$	Isothermal particle	$0.001 < St_{v,cr} < 10$	$20\,\mu m < D < 1500\,\mu m$
$St_v \gg 1$	Free dropping particle	$St_{v,cr} = 10$	$D_{cr} = 1500\,\mu m$

Note The corresponding D is calculated with typical building parameters $D_T = 10$ m, H = 3 m, U = 0.1 m/s, $\rho_p = 1000$ kg/m^3

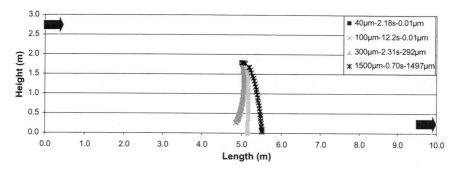

Fig. 3.11 Predicted trajectories of vaporizing particles with different sizes in the ventilated room

time the droplet diameter is barely changed. The 1500 μm droplet further exhibits the characteristics of an object in free-fall with less influence from airflow. Table 3.3 calculates the corresponding evaporation effectiveness numbers to the critical droplet diameters under typical room conditions and indicates the appropriate models for simulating droplet with different sizes.

(3) **Summary**

Different particle and droplet CFD models provide different simulation results in which the size of particle and droplet is a critical justification factor. By analyzing the particle and droplet momentum and mass conservation equations, two practical indices—the Stokes number and the Evaporation Effectiveness number are proposed to be applied as simple criteria to determine appropriate CFD models for particle and droplet prediction. The case studies provide the rules of thumb that can be used by building application engineers to guide their engineering simulations of indoor air quality under typical room conditions, as summarized in Fig. 3.12.

According to Fig. 3.12, the bacteria and viral particles can be represented fairly accurately by the lazy model because their diameters are far less than 20 μm as shown in Fig. 3.13. Bio-aerosols with nuclei that are free from evaporation, such as droplets produced during coughing or sneezing, can also be reasonably simulated by the lazy model due to their small sizes. For larger-size solid particles such as pollens and plant spores that usually have diameters of over 20 μm, the isothermal particle model may be necessary. The vaporizing droplet model is imperative for

Table 3.3 Model selection criteria for simulating vaporizing particle movement in the air

EE number	Category	Critical EE	Corresponding D
EE $\ll$ 1	Lazy particle	EE = 0.01	$D_{cr} = 40$ μm
EE $\approx$ 1	Vaporizing particle	0.01 < EE < 10	40 μm < D < 1500 μm
EE $\gg$ 1	Isothermal particle	EE = 10	$D_{cr} = 1500$ μm

Note The corresponding D is calculated with typical building parameters $D_T = 10$ m, H = 3 m, U = 0.1 m/s, $\rho_p = 1000$ kg/m^3, RH = 40%, T = 20 °C

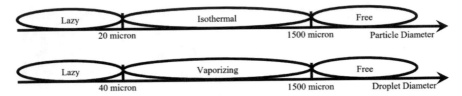

Fig. 3.12 Rules of thumb for selecting models to predict indoor particle and droplet transport

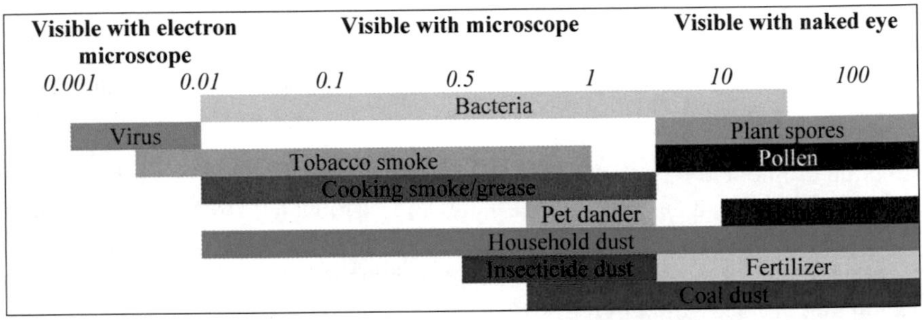

| Visible with electron microscope | | Visible with microscope | | | Visible with naked eye | |

Fig. 3.13 Typical size ranges (in micron) of indoor air pollution particles

modeling droplets from sprinklers during fire extinguishing scenarios because of the large water droplet sizes. In practice, to be safe/accurate, 5 μm is also commonly used as a critical size to judge whether a Lagrangian model is needed. Although the rules of thumb provides the initial guidance on model selection, identifying a suitable model may still require specific (and sometime iterative) investigations that consider simulation goals, computing cost and affordability, and actual particle and environment conditions.

Practice-3: Indoor Airflow and Heat Transfer

Example Project: Air distribution inside a hospital operating room (OR)

Background:

The goal of the air distribution inside a hospital operating room (OR) is to protect the patient and staff from cross-infection while maintaining occupant comfort and not affecting the facilitation of surgical tasks. However, a source of contamination bypasses HEPA installations in every OR, this source being the surgical staff themselves and the particles stirred up by their movement (Cook and Int-Hout 2009). Therefore, air motion control must be used to maximize air asepsis.

In hospital ORs, using HEPA-filtered air and vertical (downward) laminar airflow is typical. ASHRAE Standard 170-2008 (ASHRAE 2008) requires that ventilation be

Table 3.4 Laboratory experiment specifications

Room dimensions	6.1 m × 5.8 m × 2.9 m
Diffuser dimensions	2.44 m × 3.05 m
Diffuser coverage area	7.06 m²
Air change rate	31.6
Nominal face velocity	0.127 m³/s m²
Room air temperature	20 °C
Supply air temperature	18.3 °C
Room pressurization	+2.5 Pa

provided from the ceiling in a downward direction concentrated over the patient and surgical team. The area of the primary ventilation air diffusers must extend at least 305 mm beyond each side of the surgical table. It also requires that air is exhausted from at least two grilles on opposing sides of the room near the floor. It requires the use of non-aspirating, Group E outlets that provide a unidirectional flow pattern in the room (aka laminar flow diffusers). This study applied a computational fluid dynamics (CFD) tool to predict the flow pattern in a representative OR environment with standard air flow settings (Zhai and Osborne 2013).

Simulation Details:

The CFD model was built according to the full-scale laboratory experiment. The same diffuser specifications and air change rate per hour (ACH) as tested in the experiment were used in the CFD model, as well as the same room and equipment and occupant conditions, as shown in Table 3.4 and Fig. 3.14. These objects and heat gain values were chosen based on detailed on-site OR studies and measurements (Zhai et al. 2013). The equipment thermal loads as well as temperature of the patient's wound and skin can be seen in Table 3.5. Table 3.6 indicates the sizes of all of the objects in the room.

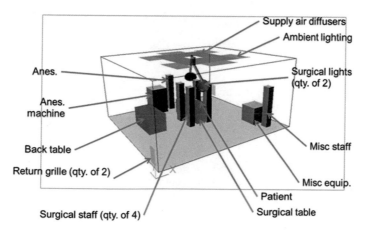

Fig. 3.14 Base CFD model setup

(a) **Geometry Generation:**

Melikov and Kaczmarczyk (2007) discussed the importance of detailed indoor objects such as human body on indoor airflow characteristics and indicated the local impacts of most details of indoor objects. Focusing on the general indoor airflow patterns and interactions between patient and medical staffs, this study simplified the simulation of indoor subjects such as human bodies and equipment as rectangular geometries (except the surgical lighting) with exact heat sources as tested. This practice facilitates the generation of high-quality meshes and therefore improves both speed and accuracy of the simulations.

(b) **Mesh Generation**

The example OR case was modeled using a rectangular Cartesian grid, which maps well to typical OR geometry. Local grid refinement was implemented near critical spaces and objects such as walls, inlets and persons. The results of a CFD simulation are highly dependent on the quality of the computational grid. The grid refinement study was conducted on the following grids: $70 \times 58 \times 45$ (180 k cells), $87 \times 73 \times 57$ (362 k cells), $106 \times 91 \times 70$ (675 k cells), $124 \times 111 \times 86$ (1.2 M cells), $155 \times 142 \times 108$ (2.4 M cells). Figure 3.15 demonstrates the finest grid distribution.

Table 3.5 Laboratory thermal loads

Object	Qty	Heat gain (W)	Temperature (°C)
Manikins—male	2	80	
Manikins—female	4	68	
Anesthesia machine	1	100	
Surgical lights	2	250	
Monitor	1	200	
Ambient lights	6	128	
Patient wound	1		25.6
Patient skin	2		27.4

Table 3.6 Room object dimensions

Object	Qty	Dimensions (m)
Surgical table	1	$0.54 \times 1.88 \times 0.66$
Back table	1	$0.76 \times 1.52 \times 0.76$
Anesthesia machine	1	$0.76 \times 0.76 \times 1.2$
Surgical lighting	2	0.58 diameter
Misc. equipment (monitor)	1	$0.76 \times 0.76 \times 0.76$
Surgical staff	6	$0.25 \times 0.30 \times 1.75$
Patient body	1	$0.30 \times 1.60 \times 0.25$

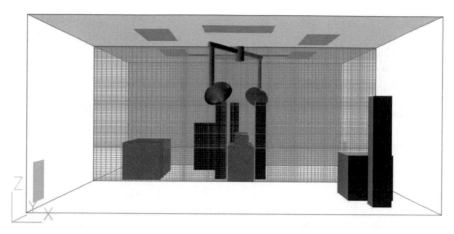

Fig. 3.15 Grid refinement case: 2.4 M cells

(c) **Solver and Models**

Both RANS and LES CFD methods were tested for this example case. While advanced CFD modeling techniques such as Large Eddy Simulation (LES) provide substantial benefits, the currently available RANS technologies have proven to be adequate for modeling the steady-state characteristics of the hospital operating room air distribution. In the RANS CFD solution methodology, the RNG $k - \varepsilon$ turbulence model (Yakhot and Orszag, 1986) was employed as suggested by Zhang et al. (2007). Details about these models will be introduced in Chap. 4 "Select Turbulence Modeling Method".

(d) **Boundary Conditions/Object Modeling**

Most indoor objects such as persons and equipment were specified straightforwardly using the standard wall/block boundary condition methods. Inlet boundary condition modeling is critical to accurate CFD modeling of indoor environments, as the inlet boundary condition is the primary source of momentum that is responsible for the overall room air distribution pattern. Srebric and Chen (2002) performed a comprehensive analysis of diffuser boundary conditions to determine appropriate simplified boundary conditions, and the box and momentum method have been determined to be the most appropriate models for the diffusers that were applied in this study. The momentum method was used in this example since it was recommended by Chen and Srebric (2002) for the grille diffuser that is similar to the non-aspirating diffuser type.

Results and Analysis:

(a) **Convergence/Grid Independence**

The simulation was considered converged when the sums of residual errors in the mass, momentum, energy, and turbulence-model equations, respectively, reach a pre-defined level (i.e., 0.1%). The grids of different sizes were evaluated using the normalized root mean squared error (NRMSE) of the CFD model results with different grids (Wang and Zhai 2012) that will be described in Chap. 9 "Analyze Results". Figure 3.17 shows the NRMSE of the predicted U and W direction velocity at the four measure poles (1–4) across the center axis of the room (2.88 m) (shown in Fig. 3.16), between the 180 K (and 362 K) meshes and the 675 K mesh. It reveals that there is generally a great improvement in error with the 362 K mesh, and the computational error is typically below 10%, and absolutely below 30%. Based on these, and in order to minimize the simulation time, the 362 K mesh could be used for various engineering parametric simulations.

(b) **Model Validation**

The simulation replicates the airflow pattern as observed in the lab experiment (Zhai et al. 2013): an inward curvature of the airflow to the center of the jet stream, as seen in Fig. 3.18. This behavior reduces the overall coverage area and could pose a contamination risk to the patient.

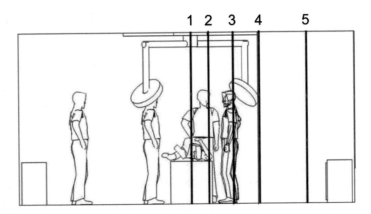

Fig. 3.16 CFD grid refinement measurement locations in central cross-sectional plane (1. center of room; 2. interior edge of diffuser; 3. midpoint of diffuser; 4. exterior edge of diffuser; 5. midpoint of outer region of room)

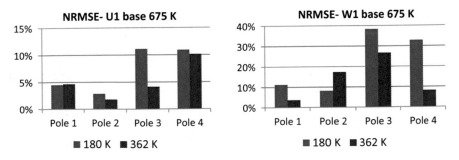

Fig. 3.17 NRMSE comparison between 180 and 362 K meshes and 675 K mesh

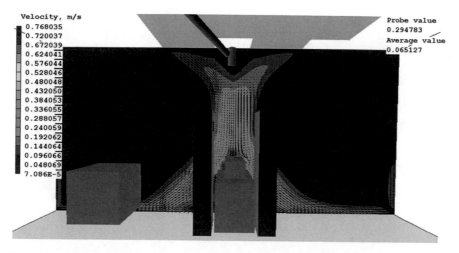

Fig. 3.18 Velocity vectors and contours at the central cross section with 675 K grid

The quantitative comparisons of simulation and experimental results were plotted in Figs. 3.19 and 3.20, for U (X) and W (Z) velocity component, respectively. Figures 3.19 and 3.20 show that the CFD simulations closely follow the experimental results, with a few exceptions (e.g., right above the patient body at Pole 1). It also appears that there is, in general, a large difference between the experimental results and the 180 K mesh, but a smaller difference between the 362 and 675 K meshes.

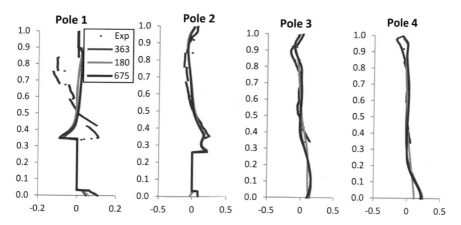

Fig. 3.19 Comparison of U-velocity in X direction

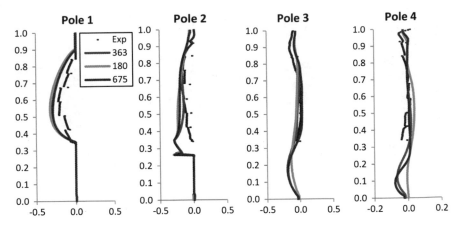

Fig. 3.20 Comparison of W-velocity in Z direction

(c) **Discussion of Results**

This example was used to demonstrate the applicability of using CFD for modeling and analysis of the surgical environment air flow. While CFD can be accurately used for modeling indoor air distribution in operating rooms, CFD user must be extremely careful in implementing these models to insure accurate simulation of air flow. The sensitivity of air flow to thermal characteristics of the indoor environment makes the model sensitive to heat gain input parameters. The heat gain and inlet boundary conditions must be carefully selected to ensure that the resulting air distribution patterns are correct.

The general indoor environment conditions place the operating room indoor air distribution in the mixed convection regime, but high cooling loads can lead to a strongly buoyancy-driven flow that is verified by the parametric study of the Archimedes number of supply air jet in the OR. The study reveals that the dependence of the room air distribution on the Archimedes number of supply air jet, instead of face velocity of supply diffuser, is of significant importance.

Assignment-3: Simulating Wind Flow Pattern across an Urban Environment

Objectives:

This assignment will use a computational fluid dynamics (CFD) program to model the wind-driven airflow distribution over an urban environment.

Key learning point:

- Urban wind simulation with appropriate domain sizes
- Wind profile input.

Simulation Steps:

(1) Build a (few) block(s) of buildings to represent a realistic community site (You may use map tools such as Google Earth to find some info);

    a. For those of you familiar with SketchUp (free tool), you may also consider to download building block models from Google SketchUp 3D warehouse for some specific real location in the world;

    b. You need to convert the SketchUp model into a certain suitable format that can be recognized and imported into the CFD software. Cleaning work is needed most of time to correctly use SketchUp models in CFD tools.

(2) Select appropriate outdoor domain sizes to be modeled;
(3) Study local weather data and identify representative wind conditions (directions, speeds, changes, frequencies, etc.);
(4) Establish corresponding boundary conditions, particularly the wind profile [isothermal case only: no temperature];
(5) Select a turbulence model: the standard k-ε model;
(6) Define convergence criterion: 0.1%;
(7) Set iteration: at least 1000 steps for steady simulation;
(8) Determine proper grid resolution with local refinement: at least 400,000 cells.

Cases to Be Simulated:

(1) Steady flow of wind over the building complex.

Report:

(1) Case descriptions: description of the case
(2) Simulation details: computational domain, grid cells, convergence status

- Figure of the grid used (on X-Z, X-Y planes);
- Figure of simulation convergence records.

(3) Result and analysis

- Figure of airflow vectors at the middle plane of the buildings;
- Figure of pressure contours at the middle plane of the buildings;
- Figure of velocity contours at the middle plane of the buildings.

(4) Conclusions (findings, result implications, CFD experience and lessons, etc.).

References

ASHRAE (2008) ANSI/ASHRAE Standard 170-2008. Ventilation of healthcare facilities. American Society of Heating, Refrigerating, and Air-Conditioning Engineers, Inc., Atlanta

Chen Q, Srebric J (2002) A procedure for verification, validation, and reporting of indoor environment CFD analyses. HVAC&R Res 8(2):201–216

Clift R, Gauvin WH (1970) The motion of particles in turbulent gas streams. In: Proceedings of Chemeca'70, vol 1, p 14

Cook G, Int-Hout D (2009) Air motion control in the hospital operating room. ASHRAE Trans 51(3):30–36

Crowe C, Sommerfeld M, Tsuji Y (1998) Multiphase flows with droplets and particles. CRC Press, Boca Raton

Jiang Y (2002) Study of natural ventilation in buildings with large eddy simulation. Ph.D. dissertation, Massachusetts Institute of Technology, Cambridge, MA, USA

Liu X, Zhai Z (2007) Identification of appropriate CFD models for simulating aerosol particle and droplet indoor transport. Indoor Built Environ 16(4):322–330 (SAGE)

Ludwig JC, Fueyo N, Malin MR (2004) The GENTRA user guide. CHAM Co

Melikov AK, Kaczmarczyk J (2007) Influence of geometry of thermal manikins on concentration distribution and personal exposure. Indoor Air 17(1):50–59

Srebric J, Chen Q (2002) Simplified numerical models for complex air supply diffusers. HVAC & R Res 8:277–294

Wang H, Zhai Z (2012) Analyzing grid-independency and numerical viscosity of computational fluid dynamics for indoor environment applications. Build Environ 52:107–118

Yakhot V, Orszag SA (1986) Renormalization group analysis of turbulence. I. Basic theory. J Sci Comput 1:3–51

Zhai Z, McNeill J, Hertzberg J (2013) Experimental investigation of hospital operating room (OR) air distribution (TRP-1397). Final report to American Society of Heating, Refrigerating, and Air Conditioning Engineers, Inc., Atlanta, 158 p

Zhai Z, Osborne A (2013) Simulation-based feasibility study of improved air conditioning systems for hospital operating room. Front Archit Res 2(4):468–475

Zhang Z, Zhang W, Zhai Z, Chen Q (2007) Evaluation of various turbulence models in predicting airflow and turbulence in enclosed environments by CFD: Part-2: comparison with experimental data from literature. HVAC&R Res 13(6):871–886

Chapter 4
Select Turbulence Modeling Method

4.1 Overview of Turbulence

Turbulence is a phenomenon of immense practical importance in nature and has therefore been extensively studied in the context of its applications, by both engineers and applied scientists. A solid grasp of turbulence, for example, can allow engineers to reduce the aerodynamic drag on an automobile or a commercial airliner, increase the maneuverability of a jet fighter, or improve the fuel efficiency of an engine. An understanding of turbulence is also necessary to comprehend the flow of blood in the heart, especially in the left ventricle, where the movement is particularly swift.

Chapter 3 introduces the governing equations for fluid flows. The general governing equations finally obtained for incompressible fluid flows, are

$$\frac{\partial u_j}{\partial x_j} = 0 \tag{4.1}$$

$$\frac{\partial u_i}{\partial t} + \frac{\partial u_j u_i}{\partial x_j} = -\frac{\partial p}{\partial x_i} + \frac{\mu}{\rho}\frac{\partial^2 u_i}{\partial x_j \partial x_j} - g_i \beta (T - T_\infty) \tag{4.2}$$

$$\frac{\partial T}{\partial t} + \frac{\partial u_j T}{\partial x_j} = \frac{1}{\rho c_p}\frac{\partial}{\partial x_k}\left(\kappa \frac{\partial T}{\partial x_k}\right) + \frac{q_{else}}{c_p} \tag{4.3}$$

If the species concentration, such as those of moisture and pollutants, are concerned in the flow domain, the concentration transport equation can be attained, in a very similar form as the energy equation:

$$\frac{\partial C}{\partial t} + \frac{\partial u_j C}{\partial x_j} = \frac{\partial}{\partial x_k}\left(\alpha \frac{\partial C}{\partial x_k}\right) + q_{source} \tag{4.4}$$

where, C is the instantaneous species concentration, α is the molecular diffusion coefficient for the species, and q_{source} is the species source.

© Springer Nature Singapore Pte Ltd. 2020
Z. Zhai, *Computational Fluid Dynamics for Built and Natural Environments*, https://doi.org/10.1007/978-981-32-9820-0_4

Equations (4.1)–(4.4) provide the instantaneous information of fluid flows such as velocity, pressure, temperature, and concentration. These equations are highly non-linear and self-coupled. It is impossible to obtain analytical solutions of these equations for most real flow problems. Therefore, the governing equations of fluid flows must be solved numerically. The numerical solution provides the spatially (and temporally) discrete information of pressure, velocity, temperature, and moisture or contaminant concentrations. This is called the *Computational Fluid Dynamics (CFD)* technique.

The flows represented by Eqs. (4.1)–(4.4) can be laminar, turbulent, or transitional between laminar and turbulent flows. Turbulence is characterized as chaotic state of fluid motion, which most real flows are. As yet no complete theory on turbulence exists, because its nonlinear dynamics are not well understood. Due to the sophisticated characteristic properties of turbulence, such as, irregularity, nonlinearity, diffusivity, large Reynolds numbers, three-dimensional vorticity fluctuations, dissipation and continuum (Tennekes and Lumley 1972), it is difficult to identify whether airflow in a room, for instance, is a locally artificially induced, transitional, or fully developed turbulence. However, very few room airflows are laminar. All non-laminar room airflows could be defined as turbulence. CFD prediction on such turbulent flows is generally via three approaches:

- Direct Numerical Simulation (DNS),
- Large Eddy Simulation (LES), and
- Reynolds-Averaged Navier-Stokes (RANS) equation simulation with turbulence models.

Figure 4.1 illustrates the transition of large-scale eddies into dissipating eddies, as well as the eddy scales that DNS, LES and RANS can, respectively, handle directly

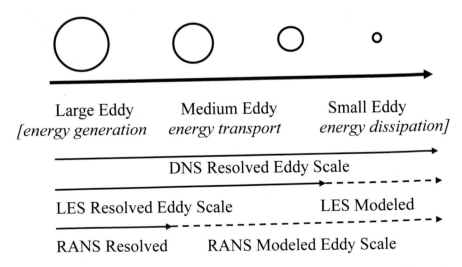

Fig. 4.1 Illustration of the transition of large-scale eddies into dissipating eddies, and the scales that DNS, LES and RANS can, respectively, handle directly (resolved) and indirectly (modeled)

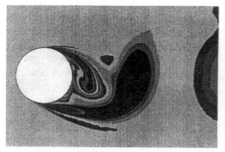
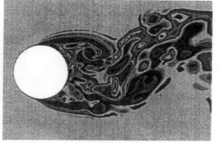

(a) Unsteady RANS model (Shur et al., 1996)   (b) LES (Travin et al., 2000)

Fig. 4.2 Simulation of flow past a circular cylinder by two CFD turbulence modeling methods

Table 4.1 Comparison of estimated grids, iterations and computing costs required for different turbulence modeling methods

Name	Unsteady	Empiricism	Grid		Iteration Step		Computing cost	
			Building $Re \sim 10^6$	Airline $Re \sim 10^8$	Building $Re \sim 10^6$	Airline $Re \sim 10^8$	Building $Re \sim 10^6$	Airline $Re \sim 10^8$
RANS	No	Strong	10^5	10^7	10^3	10^3	2 h	20 h
Unsteady RANS	Yes	Strong	10^5	10^7	$10^{3.5}$	$10^{3.5}$	6 h	60 h
LES[a]	Yes	Slightly	10^6	10^8	10^4	10^4	3 days	9 days
LES[b]	Yes	Weak	10^8	$10^{11.5}$	$10^{5.5}$	$10^{6.7}$	7 days	25 days
DNS	Yes	None	10^{11}	10^{16}	10^6	$10^{7.7}$	20 days	2 months

[a]Special treatments of the boundary layer with either a wall model or a RANS model
[b]No special treatments of the boundary layer

(resolved) and indirectly (modeled). Figure 4.2 demonstrates the CFD simulation results by unsteady RANS and LES, respectively, for the flow past a circular cylinder. It is clear that the LES model presents more information on small eddies while the unsteady RANS model captures the general flow patterns. Table 4.1 compares the computational grids and iterations required by different turbulence modeling methods for both building and aircraft applications, as well as the estimated computing costs on a modern personal computer.

4.2 Direct Numerical Simulation (DNS)

DNS computes a turbulent flow by directly solving the highly reliable Navier-Stokes equation without approximations. DNS resolves the whole range of spatial and temporal scales of the turbulence, from the smallest dissipative scales (Kolmogorov scales) to the integral scale, L (case characteristic length), which is associated with the motions containing most of the kinetic energy. As a result, DNS requires a very fine grid resolution to capture the smallest eddies in the turbulent flow. According to turbulence theory (Nieuwstadt 1990), the number of grid points required to describe turbulent motions should be at least N ~ $Re^{9/4}$. The computer systems must become rather large (memory at least 10^{10} words and peak performances at least 10^{12} flops) in order to conduct computations for the flow (Nieuwstadt et al. 1994). In other words, since the smallest eddy size is about 0.01–0.001 m, at least 1000 × 1000 × 1000 grids are needed to solve airflow in a room, for instance. This requires a very high-end computer system such as a super computer with possible parallel computing to perform a simulation. In addition, the DNS method requires very small time steps, which makes the simulation extremely time consuming. It is generally agreed that applying DNS for most real applications is still not feasible now or in the near future.

4.3 Large Eddy Simulation (LES)

According to the Kolmogorov's theory of self similarity (Kolmogorov 1941), large eddies of turbulent flows are dependent on the geometry while the smaller scales more universal. Smagorinsky (1963) and Deardorff (1970) developed a method named as "Large-Eddy Simulation" with the hypothesis that the turbulent motion could be separated into large-eddies and small-eddies such that the separation between the two does not have a significant effect on the evolution of large-eddies. The large-eddies corresponding to the three-dimensional, time-dependent equations can be directly simulated in computer without approximations. Turbulent transport approximations are then made for small-eddies independently from the flow geometry, which eliminates the need for a very fine spatial grid and a small time step. The philosophy behind this approach is that the macroscopic structure is characteristic for a turbulent flow. Moreover, the large scales of motion are primarily responsible for all transport processes, such as the exchange of momentum and heat. The success of the method stems from the fact that the main contribution to turbulent transport comes from the large-eddy motion. Hence, the large-eddy simulation is clearly superior to turbulent transport closure wherein the transport terms (e.g., Reynolds stresses, turbulent heat fluxes, etc.) are treated with full empiricism. In the last decades, rapid advances in computer capacity and speed have made it possible to use LES for industrial applications. Many successful case studies can be found in literature for various applications of different scales and natures. LES provides detailed information of instantaneous fluid flow and turbulence with the compensation of still considerable computing time. One of the recent efforts to reduce this computing cost while keeping the simulation precision is to combine LES with RANS, named as Detached Eddy Simulation, which will be introduced later.

4.4 Reynolds-Averaged Navier-Stokes (RANS) Equations

For most industrial applications, such as air distribution in an enclosed environment, the mean flow parameters (e.g., velocity, pressure, temperature) are more useful and important than instantaneous turbulent flow parameters (e.g., turbulence intensity). Therefore, the interest is stronger in solving the RANS equations with turbulence models that can quickly predict macro flow characteristics. The RANS approach calculates statistically averaged (Reynolds-averaged) variables for both steady-state and dynamic flows and simulates turbulence fluctuation effects on the mean fluid flows by using different turbulence models. Many turbulence models have been developed since the 1970s but a generic turbulence model that is suitable for all flows is not yet available (likely will never be available due to the inherent complications of turbulence for different flow conditions). A few prevalent turbulence models were developed, adopted and verified for engineering applications, such as the standard k-ε model (Launder and Spalding 1974). Despite the challenges associated with turbulence modeling, the RANS approach has become very popular in modeling fluid engineering problems due to its significantly small requirements on computer resources and user skills as well as reasonable accuracy.

Reynolds (1895) decomposed the instantaneous velocity and pressure and other variables into a statistically averaged value (denoted with capital letters) and a turbulent frustration superimposed thereon (denoted with $'$ superscript):

$$u_i = U_i + u_i' \quad p = P + p' \quad \phi = \Phi + \phi'$$ (4.5)

And the average operation on these instantaneous, averaged and fluctuant variables follows the general Reynolds average rules. Taking velocity as an example, the Reynolds average rules can be summarized as:

$$\overline{u_i} = \overline{U_i} = U_i \quad \overline{u_i'} = 0$$ (4.6a)

$$\overline{u_i' U_j} = 0 \quad \overline{u_i + u_j} = U_i + U_j$$ (4.6b)

$$\overline{u_i u_j} = U_i U_j + \overline{u_i' u_j'}$$ (4.6c)

When operating the Reynolds average on all the terms in Eqs. (4.1)–(4.4), the Reynolds-Averaged Navier-Stokes (RANS) Equations for incompressible fluid flows can be obtained:

• Continuity equation

$$\frac{\partial U_i}{\partial x_i} = \frac{\partial u_i'}{\partial x_i} = 0$$ (4.7)

- Momentum equations

$$\frac{\partial U_i}{\partial t} + U_j \frac{\partial U_i}{\partial x_j} = -\frac{\partial P}{\partial x_i} + \frac{\partial}{\partial x_j}\left(\nu\frac{\partial U_i}{\partial x_j} - \overline{u_i'u_j'}\right) - g_i\beta(T - T_\infty) \qquad (4.8)$$

- Energy equation

$$\frac{\partial T}{\partial t} + U_j \frac{\partial T}{\partial x_j} = \frac{\partial}{\partial x_k}\left(\Gamma\frac{\partial T}{\partial x_k} - \overline{u_k'T'}\right) + q_{source} \qquad (4.9)$$

- Concentration equation

$$\frac{\partial C}{\partial t} + U_j \frac{\partial C}{\partial x_j} = \frac{\partial}{\partial x_k}\left(\alpha\frac{\partial C}{\partial x_k} - \overline{u_k'c'}\right) + q_{source} \qquad (4.10)$$

where, $\nu = \frac{\mu}{\rho}$ is the molecular kinematic viscosity, $\Gamma = \frac{k}{\rho c_p} = \frac{\nu}{Pr}$ is the temperature viscous diffusion coefficient and Pr is the Prandtl number, $\alpha = \frac{\nu}{Sc}$ is the concentration viscous diffusion coefficient and Sc is the Schmidt number. $\overline{u_i'u_j'}$, $\overline{u_k'T'}$ and $\overline{u_k'c'}$, called as Reynolds stresses, turbulent heat flux, and turbulent concentration flux, respectively, are unknown. They represent the turbulence influence on the mean flow, temperature and concentration development, which need to be solved or modeled approximately.

4.5 Turbulence Models

In the last hundred years, many turbulence models have been developed to represent the unknown $\overline{u_i'u_j'}$, $\overline{u_k'T'}$, $\overline{u_k'c'}$ so as to enclose Eqs. (4.7)–(4.10). These turbulence models can be generally divided into two categories: eddy-viscosity models and Reynolds stress models. Figure 4.3 illustrates an incomplete overview of various turbulence models.

(1) Eddy-Viscosity Models

The eddy-viscosity models adopt Boussinesq approximation (1877) that relates Reynolds stress to the rate of mean stream through an "eddy" viscosity.

$$\overline{u_i'u_j'} = \frac{2}{3}\delta_{ij}k - \nu_t\left(\frac{\partial U_i}{\partial x_j} + \frac{\partial U_j}{\partial x_i}\right) \qquad (4.11)$$

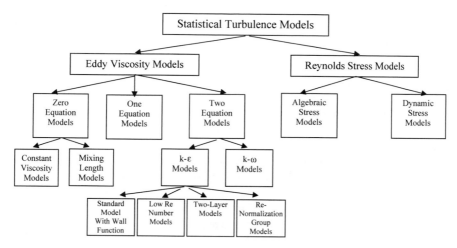

Fig. 4.3 Overview of turbulence models

Similarly, the turbulent scalar fluxes, such as turbulent heat and concentration fluxes, can be approximated as additional diffusions caused by turbulence (eddy-diffusivity)

$$\overline{u_k' T'} = -\frac{v_t}{Pr_t} \frac{\partial T}{\partial x_k} \tag{4.12}$$

$$\overline{u_k' c'} = -\frac{v_t}{Sc_t} \frac{\partial C}{\partial x_k} \tag{4.13}$$

where δ_{ij} is the Kronecker delta (when $i \neq j$, $\delta_{ij} = 0$; and when $i = j$, $\delta_{ij} = 1$), and k is the turbulence kinetic energy ($k = \frac{\overline{u_i' u_i'}}{2}$). Pr_t is the turbulent Prandtl number, and Sc_t is the turbulent Schmidt number.

How to obtain this "eddy" viscosity v_t is the core of conventional eddy-viscosity model development. Eddy-viscosity models are normally classified according to the number of transport equations used. Classic eddy-viscosity models include mixing-length model (zero equation eddy-viscosity model) (e.g., Prandtl 1925), one equation eddy-viscosity model (e.g., Kolmogorov 1942), two equations eddy-viscosity model (e.g., Launder and Spalding 1974). The following sections review various eddy-viscosity models from the simplest to the most complex ones.

- *Zero-equation eddy-viscosity models*

The zero-equation turbulence eddy-viscosity models are the simplest eddy viscosity models. The model has one algebra equation for turbulent viscosity, and no (zero) additional partial differential equations (PDE) beyond the Reynolds-averaged equations for mass, momentum, energy and species conservation.

The earliest zero-equation model was developed by Prandtl (1925) with the mixing-length hypothesis. In analogy to the molecular transport with fluid lump motions, the mixing-length model assumes:

$$v_t = L_{mix}^2 \left| \frac{dU}{dy} \right| \tag{4.14}$$

where L_{mix} is the mixing length that needs to be prescribed. Although the mixing-length model is not theoretically sound and L_{mix} need calibrations for each specific type of flows (sometimes it is difficult), the model has yielded plausible results in predicting simple turbulent flows such as free shear flows.

Many efforts have been made to extend the applicability of the mixing-length model, such as the contributions by Van Driest (1956), Cebeci and Smith (1974), Baldwin and Lomax (1978). Xu (1998) used the assumption of uniform turbulence intensity and derived an algebraic function to express turbulent viscosity as a function of local mean velocity, U, and the distance to the nearest wall, L:

$$v_t = 0.03874 \, U \, L \tag{4.15}$$

The equation has an empirical constant, a universal value of 0.03874 for different flows. The validations conducted by Chen and Xu (1998), Srebric et al. (1999), Morrison (2000) have demonstrated the feasibility of this model in predicting room airflows. In fact, Li et al. (2005) applied this zero-equation model for outdoor thermal environment simulations, which also provided reasonable predictions when compared with the measured data.

Although zero-equation models have fatal physical deficiencies, for instance, without considering non-local and flow-history effects in the eddy-viscosity, and although more sophisticated turbulence models are developed, zero-equation models still gain certain attentions in today's industrial practices because they are simple, cost-effective, and once calibrated, can predict mean-flow quantities fairly well. In fact, some simple zero-equation models may provide surprisingly good results. For example, a constant viscosity model (an empirical constant v_t) can give much better results for swirling flow than the standard k-ε model. In addition, Nielsen's study (1998) showed that the constant eddy-viscosity model provides results closer to the measured data than the standard k-ε model for the prediction of smoke movement in a tunnel. Nilsson (2007) also used the constant eddy-viscosity model to study the comfort conditions around a thermal manikin, which provided acceptable accuracy with significantly less computing efforts.

• *One-equation eddy-viscosity models*

The turbulent viscosity correlations of zero-equation models often fail due to the inherent physical deficiencies such as not considering non-local and flow-history effects on turbulent eddy-viscosity. One-equation turbulence models use additional turbulence variables (e.g., the turbulent kinetic energy $k = \frac{1}{2}\overline{u_i'u_i'}$) to calculate eddy viscosity v_t, such as:

$$v_t = Ck^{1/2}l \tag{4.16}$$

where k is obtained by solving a transport equation, l is a turbulence length scale, and C is a constant coefficient. The one-equation models need to prescribe the length scale l in a similar manner as that for the zero-equation models.

Most one-equation models solve the transport equation for turbulent kinetic energy k. Some one-equation models derive transport equations for other turbulent variables, such as the turbulent Reynolds number (Baldwin and Barth 1990). Spallart and Allmaras (1992) proposed to directly solve a transport equation for eddy viscosity (the S-A model). Unlike most other one-equation models, the S-A model is local so that the solution at one point is independent of the solutions at neighboring cells and thus compatible with grids of any structure. This model is most accurate for free shear and boundary layer flows. Literature review shows that the S-A model, among very few one-equation models used for confined space simulation, is a relatively popular and reliable one-equation model. Toraño et al. (2006) simulated ventilation in tunnels and galleries with the constant turbulent eddy viscosity, the k-ε, and the S-A models. The comparison of simulation results with detailed experimental data showed great performance of the k-ε and the S-A models. In addition, the S-A model has been typically incorporated by one of the newest turbulence modeling methods—detached eddy simulation (DES), which will be discussed later.

- *Two-equation eddy-viscosity models*

In addition to the k-equation, two-equation eddy-viscosity models solve a second partial differential transport equation for z ($z = k^\alpha l^\beta$) to represent more turbulence physics. Different α and β values form various kinds of two-equation models. Two-equation models are generally superior to zero- and one-equation models because they do not need prior knowledge of turbulence structures. The eddy viscosity can be calculated from the k and the length scale, l. Table 4.2 lists some typical two-equation models.

(a) *k-ε two equations eddy-viscosity models*

The k-ε model family is the most popular turbulence model and has the largest number of variants. By solving two additional partial differential equations to acquire parameters for the calculation of turbulent viscosity, two-equation eddy-viscosity models take into account more physics of turbulence. The k-ε two equations eddy-viscosity model developed by Launder and Spalding in 1974, so-called the "standard" k-ε model, has been widely used in practice, where k is turbulence kinetic energy and ε is the dissipation rate of turbulence energy. The eddy viscosity ν_t is defined as:

Table 4.2 Typical forms of z variable in two-equation eddy-viscosity models (Zhai et al. 2007)

z	$k^{1/2}/l$	$k^{3/2}/l$	k/l^2	k/l
Symbol	ω	ε	W	kl
Reference	Kolmogorov (1942)	Chou (1945)	Spalding (1972)	Rodi and Spalding (1984)

$$v_t = C_\mu \frac{k^2}{\varepsilon} \tag{4.17}$$

where $C_\mu = 0.09$ is the empirical constant. k and ε can be determined by solving the transport equations of k and ε:

$$U_j \frac{\partial k}{\partial x_j} = \frac{\partial}{\partial x_j}\left[\left(v + \frac{v_t}{\sigma_k}\right)\frac{\partial k}{\partial x_j}\right] + P + G - \varepsilon \tag{4.18}$$

$$U_j \frac{\partial \varepsilon}{\partial x_j} = \frac{\partial}{\partial x_j}\left[\left(v + \frac{v_t}{\sigma_\varepsilon}\right)\frac{\partial \varepsilon}{\partial x_j}\right] + [C_{\varepsilon 1}(P + G) - C_{\varepsilon 2}\varepsilon]\frac{\varepsilon}{k} \tag{4.19}$$

where,

$$P = v_t \frac{1}{2}\left(\frac{\partial U_i}{\partial x_j} + \frac{\partial U_j}{\partial x_i}\right)^2, G = -g_k\beta\frac{v_t}{\sigma_t}\frac{\partial T}{\partial x_k} \tag{4.20}$$

and $\sigma_t = 1.0$, $\sigma_k = 1.0$, $\sigma_\varepsilon = 1.3$, $C_{\varepsilon 1} = 1.44$, $C_{\varepsilon 2} = 1.92$ are the empirical constants.

Numerous other two-equation models have been suggested afterwards. Chen (1995b) has tested five different k-ε models for natural convection, forced convection, mixed convection and impinging jet in a room, but it is very difficult to identify any other models superior to the standard k-ε model.

Re-Normalization Group (RNG) k-ε model (Yakhot and Orszag 1986) has also gained popularity for predicting engineering flows with many successes. The RNG-based k-ε turbulence model is derived from the instantaneous Navier-Stokes equations, using a statistical technique called "renormalization group" (RNG) theory. The analytical derivation results in a model with constants different from those in the standard k-ε model, as well as additional terms and functions in the transport equations for k and ε. The RNG model has an additional term in its ε equation that improves the accuracy for rapidly strained flows. The effect of swirl on turbulence is included in the RNG model, enhancing accuracy for swirling flows. In addition, the RNG theory provides an analytical formula for turbulent Prandtl numbers, while the standard k-ε model uses user-specified, constant values. Furthermore, the RNG theory provides an analytically-derived differential formula for effective viscosity that accounts for low-Reynolds number effects. Effective use of this feature does, however, depend on an appropriate treatment of the near-wall region.

Another high Reynolds number k-ε model family is realizable k-ε models, which derive the transport equations of k-ε model with physics-based realizable constraints and rules. The term "realizable" means that the model satisfies certain mathematical constraints on the Reynolds stresses, consistent with the physics of turbulent flows. Neither the standard k-ε model nor the RNG k-ε model is realizable. Realizable k-ε models usually provide much improved results for swirling flows and flows involving separation when compared to the standard k-ε model. For example, Van Maele and

Merci (2006) indicated that the realizable k-ε model (Shih et al. 1995) performs better than the standard k-ε model for predicting various buoyancy plumes.

The high Reynolds number models such as the standard k-ε model usually fail when being applied for low Reynolds number flows such as near-wall regions. When using two-equation eddy-viscosity models to simulate flows with rigid boundaries, such as, walls in a room, special flow-handling approach need to be employed in the wall's vicinity where the presence of a wall significantly reduces the turbulence level. The standard wall function approach (Launder and Spalding 1974) are the most widely used approximation in practice. The wall functions connect the outer-wall free stream and the solid walls by a set of prescribed functions. The use of wall functions avoids modeling the rapid changes of flow and turbulence near the walls with a fine grid and thus saves the computing time. Another method that can be used to determine the wall effects is two-layer models (e.g., Rodi 1991). The two-layer models divide the wall vicinity into a viscosity-affected near-wall region resolved with a one-equation model and an outer region simulated with the standard k-ε model.

Another approach for handling near-wall flows is to use a low-Reynolds-number (LRN) turbulence model to solve the governing equations all the way down to the solid surfaces. LRN models request very fine grid near the walls so that the computing time is much longer. Tens of LRN models have been proposed since the 1970s while most of them have the similar form. By popularity, it was sensed that the LRN models developed by Jones and Launder (1973), Launder and Sharma (1974) are the most commonly used models, upon which a few variation models were developed (e.g., Radmehr and Patankar 2001) but less used. The observation of the applications of LRN models for indoor simulation reveals that the LRN model may only improve model accuracy for specific cases and lack wide applicability.

Chen (1995b) compared five k-ε based turbulence models in predicting various convective airflows and an impinging flow. The results showed that the RNG k-ε model had the best overall performance in terms of accuracy, numerical stability, and computing time, while the standard k-ε model had competitive performance. The findings were confirmed by many following studies.

(b) *k-ω two-equation eddy-viscosity model*

The k-ω two-equation eddy-viscosity models (e.g., Wilcox 1988; Menter 1994) have also received increasing attentions in many industrial applications in the last decades. In the k-ω models, ω is the ratio of ε over k. Compared to the k-ε models, the k-ω models are superior in predicting equilibrium adverse pressure flows (Wilcox 1988; Huang et al. 1992) while less robust in wake region and free shear flows (Menter 1992). This led to the development of an integrated model that takes advantages of both models, a fairly successful model named shear stress transport (SST) k-ω model developed by Menter (1994). The SST k-ω model is essentially a k-ω model near wall boundaries while is equivalent to a transformed k-ε model in regions far from walls. The switch between the k-ω and k-ε formulations is controlled by blending functions.

The k-ω models present a new potential to model fluid flows with good accuracy and numerical stability. A few existing studies indicate that the SST k-ω model has a better overall performance than the standard k-ε model and the RNG k-ε model. Recently, one of the commercial CFD software, CFX, has placed its emphasis on ω-equation based turbulence models due to its multiple advantages, such as simple and robust formulation, accurate and robust wall treatment (low-Re formulation), high quality for heat transfer predictions, and easy combination with other models.

- *Multiple-equation eddy-viscosity models*

Another noticed development in eddy-viscosity models is multiple-equation eddy-viscosity models. A multiple-equation eddy-viscosity model is often developed and used for near-wall flows. Durbin (1991) suggested that the wall blocking effect, i.e., zero normal velocity at walls, is much more crucial than the viscous effect on near-wall flows. Instead of using the turbulent kinetic energy to calculate near-wall turbulence eddy viscosity, he suggested the use of a more proper quantity, the fluctuation of normal velocity $\overline{v'^2}$, as the velocity scale in the near-wall eddy viscosity calculation. Durbin introduced a transport equation of $\overline{v'^2}$ and a corresponding damping function f for the $\overline{v'^2}$ equation, which thus created a three-equation eddy-viscosity model (named v2f model) including k, ε and $\overline{v'^2}$ transport equations. The model received continuous improvement and modification afterwards (e.g., Durbin 1995; Lien and Durbin 1996; Davidson et al. 2003; Laurence et al. 2004).

The v2f model, as one of the most recently developed eddy viscosity models, has a more solid theoretical ground than LRN models but is less stable for segregated solvers. Choi et al. (2004) tested the accuracy and numerical stability of the original v2f model (Durbin 1995) and a modified v2f model (Lien and Kalitzin 2001) along with a two-layer model (Chen and Patel 1988) for natural convection in a rectangular cavity. The study found the original v2f model with the algebraic heat flux model best predicted the mean velocity, velocity fluctuation, Reynolds shear stress, turbulent heat flux, local Nusselt number, and wall shear stress. The predicted results agreed fairly well with the measurements. However, this model exhibits the numerical stiffness problem in a segregate solution procedure such as the SIMPLE algorithm, which requires remedy. Davidson et al. (2003) discovered that the v2f model could over-predict $\overline{v'^2}$ in regions far away from walls. They, therefore, analyzed the f equation in isotropic condition and postulated a simple but effective way to limit $\overline{v'^2}$ in nearly isotropic flow regions. With this restriction function, the v2f model can improve the accuracy in regions far away from walls. The v2f model brings more turbulence physics especially for low speed near-wall flows, which are critical in enclosed environments.

Other than the v2f models, some other multiple-equation eddy-viscosity models can be found in literature. For instance, Hanjalic et al. (1996) proposed a new three-equation eddy-viscosity model by introducing a transport equation for RMS temperature fluctuation $\overline{\theta'^2}$ for high Raleigh number flows. However, all these models become more complicated and have not been well accepted and applied for predicting complex flow conditions.

(2)   **RANS Reynolds Stress Models**

Most eddy viscosity models assume isotropic turbulence structures, which could fail for flows with strong anisotropic behaviors, such as swirling flows and flows with strong curvatures. Reynolds stress models (RSM), instead of calculating turbulence eddy viscosity, explicitly solve the transport equations of Reynolds stresses and fluxes. However, the derivation of the Reynolds stresses transport equations leads to higher order unsolved turbulence correlations, such as $\overline{u'_i u'_j u'_k}$, which need be modeled to close the equations.

The development and application of Reynolds stress models can be traced back to the 1970s. The applications of Reynolds stress models in the early 1980s were mainly for thin shear flow such as Hah and Lakshminarayana (1980), Gibson and Rodi (1981), Hossain and Rodi (1982). Studies directed toward three-dimensional flows began to appear in the 1990s. Early applications of RSM in room airflow computation include those by Murakami et al. (1990) and Renz and Terhaag (1990). They computed airflow patterns in a room with jets. The results showed that the Reynolds stress model is superior to the standard k-ε model, because anisotropic effects of turbulence are taken into account. The same conclusions were reached by Moureh and Flick (2003) who investigated the characteristic of airflow generated by a wall jet within a long and empty slot-ventilated enclosure. Dol and Hanjalic (2001) predicted the turbulent natural convection in a side-heated near-cubic enclosure. They found that the second-moment closure is better in capturing thermal three-dimensionality effects and strong streamline curvature in the corners while the k-ε model still provides reasonable predictions of the first moments away from the corners.

Chen (1996) compared three Reynolds-stress models with the standard k-ε model for natural convection, forced convection, mixed convection, and impinging jet in a room. He concluded that the Reynolds-stress models are only slightly better than the k-ε model but have a severe penalty in computing time. Based on a large number of applications for engineering flows, Leschziner (1990) concluded that Reynolds stress models is appropriate and beneficial when the flow is dominated by a recirculation zone driven by a shear layer. Among various RSMs, the model developed by Gibson and Launder (1978) and the one by Gatski and Speziale (1993) are often used in practice. The models, however, still have some weaknesses that need to be addressed. Tornstrom and Moshfegh (2006), for instance, found that the RSM with linear pressure-strain approximation over-predicted the lateral spreading rate and the turbulent quantities of 3-D cold wall jets.

In general, the applications of Reynolds stress models need (three to ten times) more computing time than those with eddy-viscosity models because of greater algebraic complexity. Meanwhile, there are still defects and weakness in Reynolds stress models needed to be solved (Launder 1989). The models need significant justification of application advantages before they can be soundly accepted and used for engineering flow predictions. Most existing practical studies indicated that the marginal improvement on prediction quality of RSM is not well justified by the high computational costs.

To reduce the computing time of the RSMs, algebraic Reynolds stress models (ASM) were developed (e.g., Rodi 1976) accordingly. The ASM derives algebraic equations for all Reynolds stress and fluxes from the differential stress models, in which each Reynolds stress correlates with others and the derivatives of velocity and temperature. Jouvray et al. (2007) tested several ASMs against the standard k-ε and k-l models (Wolfshtein 1969) for two rooms with displacement and mixing ventilation. The results showed that the nonlinear ASMs gave marginally better agreement with the measured data than did others. The application of the nonlinear models may not be well justified due to the strong case-dependent stability performance and the high additional computational costs.

(3) Large Eddy Simulation Models

LES models have been receiving increased attentions for modeling engineering flows due to the rapid development of computer speed and power. LES is an intermediate modeling technique between DNS and RANS. LES solves filtered (transformed) Navier-Stokes equations for large-scale eddies while models small-scale (also known as subgrid scale) eddies. Filtering of various variables in the Navier-Stokes equations is similar to the process of Reynolds averaging and the resulting equations for incompressible flow can be written in a similar form as the RANS equations. Smagorinsky (1963) proposed the first subgrid model correlating eddy viscosity to the strain rate, which can be written in the form of eddy viscosity as follows,

$$\tau_{ij} = \frac{1}{3}\tau_{kk}\delta_{ij} - 2\upsilon_t \overline{S}_{ij} \tag{4.21}$$

where, $\overline{S}_{ij}$ is the strain rate tensor based on the filtered velocity field, the isotropic part τ_{kk} is a unknown scalar and is usually combined with $\bar{p}$. The eddy viscosity is expressed as

$$\upsilon_t = (c_s \Delta)^2 |\overline{S}| \tag{4.22}$$

where, $|\overline{S}| = \sqrt{2\overline{S}_{ij}\overline{S}_{ij}}$, Δ is the filter width and C_s is the Smagorinsky constant. Different Smagorinsky constants C_s were proposed by various researchers. Lilly (1966) suggested a value of 0.17 for C_s in homogeneous isotropic turbulent flow. Many variants of Smagorinsky model were proposed thereafter. In physics, the C_s may not be a constant. Thus, the dynamic Smagorinsky-Lilly model based on the Germano identity (Germano et al. 1991; Lilly 1992) calculates the C_s with the information from resolved scales of motion

$$(C_s)^2 = \frac{\langle L_{ij}M_{ij}\rangle}{\langle M_{ij}M_{ij}\rangle} \tag{4.23}$$

where, the L_{ij} and M_{ij} are the resolved stress tensor, and < > is an average operation on homogeneous region. Without the average, the dynamic model has been found to yield a highly variable eddy viscosity field with negative values, which caused the numerical instability. However, the average operation is difficult to implement when the flow field does not have statistical homogeneity direction. Meneveau et al. (1996) proposed the Lagrangian dynamic model, in which a Lagrangian time average was applied to Eq. (4.23). Zhang and Chen (2000) proposed to apply an additional filter for Eq. (4.23), which improved the simulation of indoor airflows. Other more complex models have been proposed to improve accuracy such as the dynamic models as reviewed by Meneveau and Katz (2000).

In the last decade, LES has been growingly applied to model various fluid flows due to its rich dynamic details compared with RANS models, as well as accuracy for critical flow areas such as separation in wake zones. As indicated by most studies, the LES model provides more detailed and accurate prediction of flow distributions, which can be critical for some cases to explore the flow mechanisms. However, the high demand on computing power, time and user knowledge makes LES still mainly for research or research-level applications.

(4) Detached Eddy Simulation Models

Detached eddy simulation (DES) method presents the most recent development in turbulence modeling method, which couples the RANS and LES models to solve problems where RANS is not sufficiently accurate while LES is not affordable. The earliest DES work includes Spalart et al. (1997) and Shur et al. (1999), in which the one-equation eddy-viscosity model (Spalart and Allmaras 1992) was used for the attached boundary layer flow while LES for free shear flows away from the walls. Since the formation of eddy viscosity in RANS and LES models is similar, the S-A model and the LES model can be coupled by using this similarity. In the near wall region, the wall distance d of a cell is normally much smaller than the stream-wise and span-wise grid size. In the regions far away from the wall, the wall distance is usually much larger than the cubic root of the cell volume, Δ. Hence, the switch between the S-A model and the LES model can be determined by comparing d and Δ. When d is much larger than Δ, large eddy simulation is performed; otherwise, the RANS (S-A) model is executed.

In practice, the switch between the RANS and LES models requires more programming and computing efforts rather than simply changing the calculation of the length scale. In fact, many implementations of the DES approach allow for regions to be explicitly designated as RANS or LES regions, overruling the distance function calculation. Squires (2004) reviewed and summarized the status and perspectives of DES for aerospace applications. Keating and Piomelli (2006) combined a RANS near-wall layer with a LES outer flow with a dynamic stochastic forcing method, which can provide more accurate predictions of the mean velocity and velocity fluctuations.

Some comparison studies of DES, LES and RANS can be found in the literature such as Roy et al. (2003), Jouvray and Tucker (2005), Jouvray et al. (2007). These studies indicated that DES appears a promising model, giving the best velocity agreement and overall good agreement with measured Reynolds stresses. However, they also mentioned that the encouraging DES results could be fortuitous because the method has the potential for LES zones to occur downstream of RANS zones, and thus results in poor LES boundary conditions. In addition, the eddy resolving approaches (LES and DES) still demanded high computational costs and computer powers. As an emerging technology, DES needs more studies before it can be widely applied for engineering predictions.

4.6 Select Proper Turbulence Modeling Method

This chapter introduces the primary turbulence modeling methods that can be (and have been) applied for diverse fluid flow simulations. Although DNS sounds most accurate, it is mostly used for fundamental research and benchmark development due to its high cost. Both LES and RANS require the assistance of proper turbulence models. Many turbulence models are available (even in commercial CFD software). Each turbulence model has its own pros and cons and thus applicable range. There is no universal turbulence model that suits all conditions. The selection of a proper model depends mainly on accuracy needed and computing time afforded.

Table 4.3 presents a handful of prevalent turbulence models for predicting airflows in enclosed environments, ranging from RANS to LES. The models are organized into eight sub-categories. Based on the perceived model popularity, one prevailing turbulence model is identified for each of the eight sub-categories. Note that Table 4.3 is not a comprehensive representation of all previously developed turbulence models. Most turbulence models presented consider the turbulence at single time and length scale at a point for simplicity needed in practice. Continuously increasing knowledge of turbulence modeling will expand the pool and popular models of this selection.

It is interestingly noted that the conclusions from past studies are not always consistent. Opposite observations can be attributed to the differences in simulated cases, numerical factors (e.g., scheme, grid, and program), judging criteria, and user skills. Without knowledge of all details of the simulations and cases studied, it is difficult to pass judgment on the merits of each turbulence model based solely on the presentation in the literature. Despite the disparities among the studies in the literature, some general remarks for turbulence modeling of air distributions in enclosed environments can be stated as follows:

(1) The standard k-ε model with wall functions (Launder and Spalding 1974) is still widely used and provides acceptable results (especially for global flow and temperature patterns) with good computational economy. The model may have difficulty dealing with special room situations (e.g., high buoyancy effect and/or large temperature gradient).

(2) The RNG k-ε model (Yakhot and Orszag 1986) provides similar (or slightly better) results as the standard k-ε model and is also widely used for airflow simulations in enclosed environments.

Table 4.3 List of prevalent turbulence models for predicting airflows in enclosed environments (Zhai et al. 2007)

Model classification			Primary turbulence models used in indoor air simulations	Prevalent models identified
RANS	EVM	Zero-eqn	0-eq. (Chen and Xu 1998)	Indoor zero eq.
		Two-eqn	Standard k-ε (Launder and Spalding 1974) RNG k- ε (Yakhot and Orszag 1986) Realizable k-ε (Shih et al. 1995)	RNG k-ε
			LRN-LS (Launder and Sharma 1974) LRN-JL (Jones and Launder 1973) LRN-LB (Lam and Bremhorst 1981)	LRN-LS
			LRN k-ω (Wilcox 1994) SST k-ω (Menter 1994)	SST k-ω
		Multi-eqn	v2f-dav (Davidson et al. 2003) v2f-lau (Laurence et al. 2004)	v2f-dav
	RSM		RSM-IP (Gibson and Launder 1978) RSM-EASM (Gatski and Speziale 1993)	RSM-IP
LES			LES-Sm (Smagorinsky 1963) LES-Dyn (Germano et al. 1991; Lilly 1992) LES-Filter (Zhang and Chen 2000, 2005)	LES-Dyn
DES			DES (S-A) (Shur et al. 1999) DES (ASM) (Batten et al. 2002)	DES-SA

(3) The zero- and one-equation models with specially tuned coefficients are appropriate (sometimes even better than significantly more detailed models) for the cases with similar flow characteristics as those used to develop the models.

(4) Most LRN k-ε models and nonlinear RANS models provide no or marginal improvements on prediction accuracy but suffer from strong case-dependent stability problems and has long computing time.

(5) The Reynolds-stress models can capture some flow details that cannot be modeled by the eddy viscosity models. The marginal improvements on the mean variables, however, are not well justified by the severe penalty on computing time.

(6) The k-ω model (Wilcox 1988) presents a new potential to model airflows in enclosed environments with good accuracy and numerical stability. Most exist-

ing studies indicate that the SST k-ω model (Menter 1994) has a better overall performance than the standard k-ε and the RNG k-ε models, but a systematic evaluation (especially for modeling indoor airflows) is needed before a solid conclusion can be reached.

(7) The v2f model (Durbin 1995) looks very promising for indoor environment simulations, but needs to resolve some inherent numerical problems and undergo a comprehensive evaluation.

(8) The LES model provides more detailed and maybe more accurate predictions for indoor airflows, which could be important for understanding the flow mechanism. However, the high demand on computing time and user knowledge still makes LES a tool mainly for research and research-level applications.

(9) The DES model can be a valuable intermediate modeling approach but needs significant further study, improvement, and validation before it can be widely adopted.

Zhang et al. (2007) conducted a comprehensive evaluation on the models listed in Table 4.3 by modeling representative airflows in enclosed environments, including force convection and mixed convection in ventilated spaces, natural convection with medium temperature gradient in a tall cavity, and natural convection with large temperature gradient in a model fire room. The predicted air velocity, temperature, Reynolds stresses, and turbulent heat fluxes were compared against the experimental data. The study also compared the computing time used by each model for all the cases. The results reveal that LES provides the most detailed flow features while the computing time is much higher than RANS models and the accuracy may not always be the highest. Among the RANS models studied, the RNG k-ε and a modified v2f model have the best overall performance over four cases studied. The other models have superior performance only in some particular cases. As indicated before, each flow type favors different turbulence models due to the inherent mechanics of the models. Table 4.4 summarizes both the performance of each model in different flows and the most suitable turbulence models for each flow category.

Practice-4: Outdoor Simulation with Different Turbulence Models

Example Project: Numerical Study of Flow past a Wing-Body Junction with a Nonlinear Eddy-Viscosity Model

Background:

Wing-body junction structures (Fig. 4.4) are widely observed in fluid engineering, such as, aircraft wing-body, turbine blade-hub, and bridge pile-ground. Fluid flows across wing-body junctions exhibit sophisticated flow characteristics as illustrated in Figs. 4.4 and 4.5. In the nose region, flow separation occurs and translates into a

Table 4.4 Summary of the performance of the turbulence models tested by Zhang et al. (2007)

Cases	Compared items	Turbulence models							
		0-eq.	RNG k-ε	SST k-ω	LRN-LS	V2f-dav	RSM-IP	DES	LES
Natural convection	Mean Temp.	B	A	A	C	A	A	C	A
	Mean Velo.	D	B	A	B	A	B	D	B
	Turbulence	n/a	C	C	C	A	C	C	A
Forced convection	Mean Velo.	C	A	C	A	A	B	C	A
	Turbulence	n/a	B	C	B	B	B	C	B
Mixed convection	Mean Temp.	A	A	A	A	A	B	B	A
	Mean Velo.	A	B	B	B	A	A	B	B
	Turbulence	n/a	A	D	B	A	A	B	B
Strong buoyancy flow	Mean Temp.	A	A	A	A	A	n/c	n/a	B
	Mean Velo.	B	A	A	A	A	n/c	n/a	A
	Turbulence	n/a	C	A	B	B	n/c	n/a	B
Computing time (unit)		1	2–4		4–8		10–20	10^2–10^3	

A = good, B = acceptable, C = marginal, D = poor, n/a = not applicable, n/c = not converged

Fig. 4.4 Flow across a
wing-body junction with two
trailing vortex legs

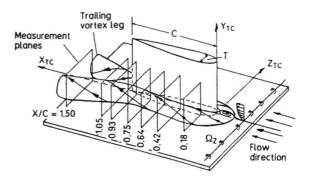

Fig. 4.5 Turbulence flow
characteristics in the nose
region around a wing-body
junction

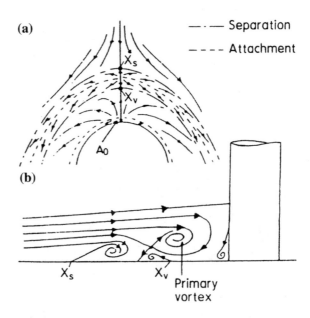

pair of horse-shoe vortices along the junction as the flow moves downstream. In the
downstream trail region, separation may occur again due to flow expansion.

Existing studies have shown that the linear eddy-viscosity models (EVM) produce excessive eddy viscosity in regions of strong surface curvature, significant
vortical motion or flow acceleration. The physical deficiency in the linear EVM can
be attributed to the isotropic assumption of the eddy viscosity, which is often incorporated in the linear stress/strain relation leading to the resultant Reynolds shear
stress direction same as the local mean velocity gradient angle. Numerous experimental studies on 3-D turbulent boundary layers have indicated that the turbulence
structure in this case, with its length and velocity scales, lags the mean flow behavior.
Mathematically, this skewness is the inequality of the flow-gradient angle, γ_g, and
the turbulent-shear-stress angle, γ_τ, which are defined as

$$\gamma_g = \tan^{-1} \frac{\partial W / \partial y}{\partial U / \partial y}, \gamma_\tau = \tan^{-1} \frac{\overline{vw}}{\overline{uv}} \qquad (4.24)$$

Obviously, $\gamma_g = \gamma_\tau$ is the direct consequence of the linear isotropic EVM.

In order to improve the predictive capability for the wing-body junction flow, progress has been achieved to employ advanced turbulence closures that can more accurately resolve the Reynolds stress anisotropy for complex 3-D flows involving large surface curvatures and strong vortical motions. In particular, the second-moment closure of SSG type (Speziale et al. 1991) has been successfully applied to such flows that clearly demonstrated the superior predictive capability in comparison with the performance of simple eddy-viscosity model.

This study attempts to explore the capability of a nonlinear EVM in this complex 3-D boundary layer prediction. The advantage of the nonlinear EVM is that it can reflect more turbulent physics than that returned by the linear EVM while greatly saving the extra amount of computation time required by the second-moment closure as compared with, for instance, $k - \varepsilon$ model computation. The present nonlinear EVM represents an explicit form of algebraic-stress model (ASM) derived by Gatrki and Speziale (1993). It therefore embodies the important stress production and redistribution mechanisms presented in the second-moment closures. The original form of the GS model however violates realizability principle in a manner of giving rise to negative turbulence energy component in many cases especially in regions of significant flow acceleration. This feature obviously represents a major defect in the original ASM and hence in the GS model. Fu et al. (1996) modified the GS model to remedy this unphysical model behavior. It was shown that the FRT nonlinear EVM (called Realizable Quadratic Eddy Viscosity Model—RQEVM) preserved the realizability property in shear flow, flow with curvature effects as well as in the flow with large distortion. This work is to evaluate the predictive quality of the RQEVM in a real complex flow (i.e., wing-body junction cross-flow), as compared with the linear EVM and GS-SSG model.

Simulation Details:

The study is a numerical investigation of the characteristics of an incompressible 3-D turbulent boundary layer generated by a wing-body junction flow. Figure 4.4 shows the experiment setting by the Simpson's group (Devenport and Simpson 1992; Fleming et al. 1993; Olcmen and Simpson 1995) and this study focuses on the case in which the wing configuration consists of a 3:2 elliptical-nose NACA 0020 tail body standing on a flat plate with T = 7.17 cm, C = 305 mm = 4.25T, and Y = 229 mm. The nominal reference velocity of the air is U_{ref} = 27.5 m/s and the Reynolds number based on T is 119,500. Figure 4.6 presents the computational domain at the dimensions of X × Y × Z = 13.4375T × 3T × 6.375T (X: flow direction; Y: perpendicular to the plate; Z: determined by the right hand rule). Only half of the flow domain was simulated due to the symmetric geometry and flow conditions. The boundary conditions adopted are marked on Fig. 4.6. The inflow boundary conditions for U, V, W, k, ε were obtained from the plate boundary layer simulation. The standard wall function was applied to both the ground (bottom)

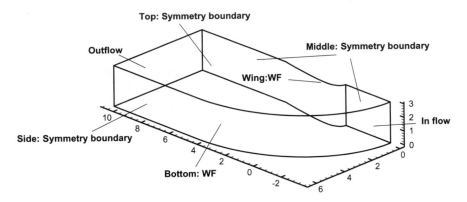

Fig. 4.6 Computational domain and boundary conditions (WF: wall function)

and the wing. Symmetry conditions were adopted for both top and side boundaries. In reality, these boundaries are close to free flow conditions. Since the focus of this study is on the corner of wing-body junction, the influences of the assumed symmetry boundary conditions on the 3-D boundary layer flows at the corner are neglectable when the domain size is adequately large. Non-uniform body-fitted structure grid was generated as shown in Fig. 4.7.

The study tested the influences of the computational domain sizes by varying the X distance from the nose (X = 0.75C, 0.8C, 1.0C), the Y height (Y = 2T, 3T, 4T), and the Z width (Z = 1.5C, 2C, 3C). No significant difference was noticed in these simulation results. Three grid systems were evaluated—X × Y × Z = 29 × 15 × 16, 60 × 36 × 43, 74 × 45 × 54—with local refinement. Considering the balance of accuracy and computing cost, the grid of 60 × 36 × 43 was used for further computations.

Results and Analysis:
Figure 4.8 illustrates the flow streamlines near the floor plate and at one cross-section at the downstream as well as the pressure on the wing surface predicted by the GS-

Fig. 4.7 Computational grid

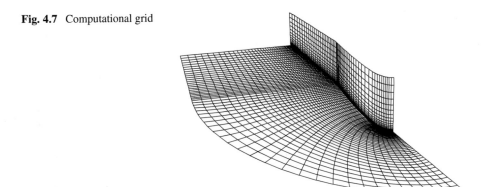

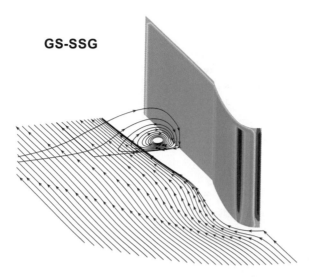

Fig. 4.8 Flow streamlines near the floor plate and at one cross-section at the downstream as well as pressure on the wing surface predicted by the GS-SSG model

SSG model. Figures 4.9 and 4.10 show the velocity vectors and streamlines of the linear EVM and nonlinear EVM results on the computational plane closest to the flat plate and on the symmetry plane ahead of the nose. It is seen that recirculation occurs in the vicinity of the nose and the flat plate. This is consistent with the oil flow visualization in the experiment and the similar computation of Chen (1995a) with the k-ε model. In Chen's work the computation with second-moment closure revealed a

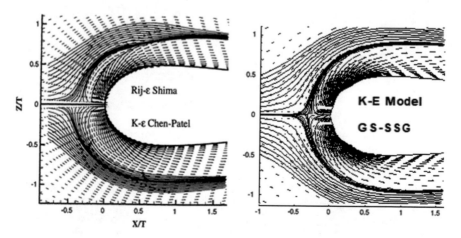

Fig. 4.9 Flow velocity vectors and streamlines of the linear EVM and nonlinear EVM results on the computational plane closest to the flat plate (left from literature add right from this study)

separation prior to the leading edge recirculation, this is however not observed here. The difference between the linear and nonlinear EVM flow patterns is not significant.

Figures 4.11 and 4.12 show the profiles of mean flow velocities and Reynolds stresses at two stations 3 and 5 representing location at $(-1.33, -2.04)$ and $(0.33, -2.94)$ in inches on X–Z plane perpendicular to the wing with the X coordinates positive to the downstream and the origin at the nose edge. All quantities in these figures have been nondimensonalized with the local free stream velocity and the maximum wing thickness. The profiles for the stream-wise velocity are almost identical among the three model calculations, but apparent difference exists between the simulation and experiment. This difference decreases as the flow moves downstream. The agreement in the span-wise velocities may be attributed partly to the uncertainties in the inlet conditions in the calculations in which the inlet profiles are taken from boundary layer calculation by matching the momentum thickness. It is not clear whether this is identical to the experimental values.

Concerning the skewness between the flow-gradient angle γ_g and the shear-stress angle γ_τ, it seems that the differences between $\overline{uv}$ and $\overline{vw}$ profiles in the linear and nonlinear EVMs represent such behaviors (Fig. 4.13). However, it can be shown analytically that in a genuine 3-D boundary layer flow, where $\partial U/\partial y$ and $\partial W/\partial y$ are the principal velocity gradients, all the quadratic nonlinear EVMs give the $\overline{uv}$ and $\overline{vw}$ shear-stress expressions the same as in the linear model. Hence, the apparent difference in the shear stresses arises due to some other secondary velocity gradients other than the 3-D boundary layer effects. The nonlinear EVMs in the quadratic form cannot truly resolve the 3-D boundary layer effects.

Unlike the mean-flow velocities, the Reynolds-stress profiles differ significantly among the computation results. While the uncertainties in the inlet profiles inevitably lead to the discrepancies in the Reynolds stresses especially at upstream locations, at the downstream locations, the nonlinear EVMs provide overall better agreement with the experiment than that returned by the linear EVM. In fact, the RQEVM results appears to be marginally better than the GS model results.

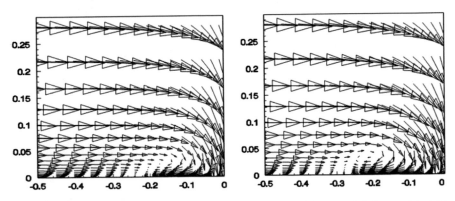

Fig. 4.10 Flow velocity vectors on the symmetry plane ahead of the nose (left: k-ε model; right: RQEVM)

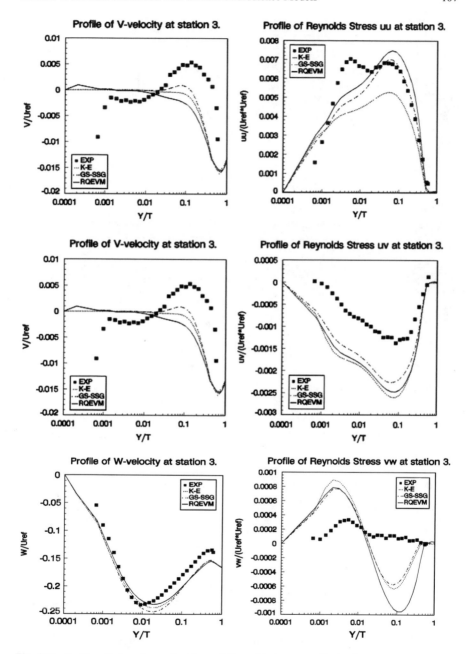

Fig. 4.11 Profiles of mean flow velocities and Reynolds stresses at Station 3

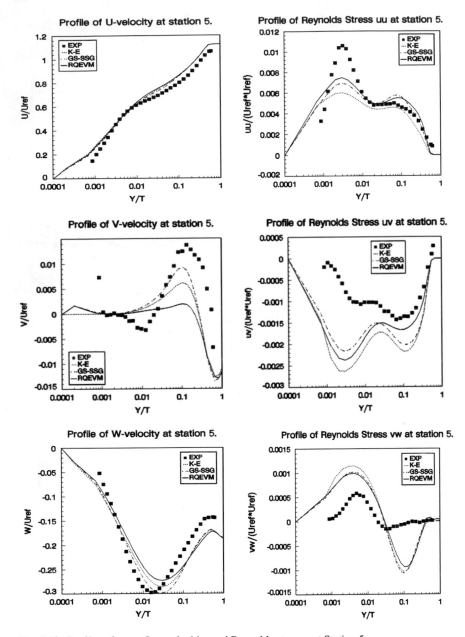

Fig. 4.12 Profiles of mean flow velocities and Reynolds stresses at Station 5

Fig. 4.13 Computed
flow-gradient angle and
shear-stress angle for Station
5

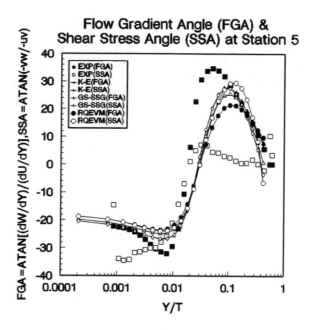

Assignment-4: Simulating Pollutant Dispersion across an Urban Environment

Objectives:

This assignment will use a computational fluid dynamics (CFD) program to model the transport of contaminant across an outdoor environment by means of natural wind.

Key learning points:

- Transient CFD simulation
- Steady and unsteady simulation of contaminant transport process.

Simulation Steps:

(1) Build a realistic but simplified outdoor environment with a set of objects (e.g., buildings, bridges, trees, highways etc.)
(2) Select appropriate outdoor domain sizes to be modeled;
(3) Study and identify representative wind conditions (directions, speeds, changes, frequencies, etc.);
(4) Select a potential (and realistic) contaminant source location and condition;
(5) Select a reasonable time period and proper time step for transient conditions;
(6) Establish corresponding boundary conditions [iso-thermal case only: no temperature];
(7) Select a turbulence model: the standard k-ε model;
(8) Define convergence criterion: 0.1%;

(9)   Set iteration: at least 1000 steps for steady simulation or 100 steps for each transient step;

(10) Determine proper grid resolution with local refinement: at least 400,000 cells.

Cases to Be Simulated:

(1)   Steady flow of wind and contaminant over the outdoor environment;

(2)   Unsteady contaminant transport over the outdoor environment under a steady wind.

Report:

(1)   Case descriptions: description of the cases

(2)   Simulation details: computational domain, grid cells, convergence status.

- Figure of the grid used (on X-Z, X-Y planes);
- Figure of simulation convergence records.

(3)   Result and analysis
      Steady conditions:

- Figure of pressure contours at the middle plane of the key objects of interest;
- Figure of velocity contours at the middle plane of the key objects of interest;
- Figure of airflow vectors at the middle plane of the key objects of interest;
- Figure of contaminant concentration contours at a proper 2-D plane of the outdoor environment;
- Figure of 3-D contaminant concentration iso-surfaces across the outdoor environment.

Unsteady conditions:

- Figures of contaminant concentration contours at a proper 2-D plane of the outdoor environment at different time steps;
- Figures of 3-D contaminant concentration iso-surfaces across the outdoor environment at different time steps.

(4)   Conclusions (findings, result implications, CFD experience and lessons, etc.)

Special Turn-In Requirements:

(1)   Animation (movie) showing the transition of the contaminant concentration contours at a proper 2-D plane of the outdoor environment at different time steps;

(2)   Animation (movie) showing the 3-D contaminant concentration iso-surfaces across the outdoor environment at different time steps.

References

Baldwin BS and Barth TJ (1990) A one-equation turbulence transport model for high reynolds number wall-bounded flows. NASA, TM-102947

Baldwin BS, Lomax H (1978) Thin-layer approximation and algebraic models for separated turbulent flows. AIAA Paper. Huntsville, AL, pp 78–257

Batten P, Goldberg U, Chakravarthy S (2002) LNS—an approach towards embedded LES. AIAA Paper: AIAA-2002-0427

Boussinesq J (1877) Essai sur la théorie des eaux courantes. Mémoires Présentés Par Divers Savants À l'Académie Des Sciences 23(1):1–680

Cebeci T, Smith AMO (1974) Analysis of turbulent boundary layers. Academic Press, New York

Chen HC (1995a) Assessment of a reynolds stress closure model for appendage-hull junction flows. J Fluids Eng 1995(117):557–563

Chen Q (1995b) Comparison of different K-E models for indoor air flow computations. Numer Heat Transfer, Part B 28:353–369

Chen Q (1996) Prediction of room air motion by reynolds-stress models. Build Environ 31(3):233–244

Chen C, Patel VC (1988) Near-Wall turbulence models for complex flows including separation. AIAA J 26:641–648

Chen Q, Xu W (1998) A zero-equation turbulence model for indoor airflow simulation. Energy Build 28(2):137–144

Choi S, Kim E, Kim S (2004) Computation of turbulent natural convection in a rectangular cavity with the K-E-V2-F model. Numer Heat Transfer, Part B 45:159–179

Chou PY (1945) On the velocity correlations and the equations of turbulent vorticity fluctuation. Quart Appl Math 1:33–54

Davidson L, Nielsen PV, Sveningsson A (2003) Modification of the v2f Model for computing the flow in a 3D wall jet. Turbulence, Heat And Mass Transfer 4:577–584

Deardorff JW (1970) A numerical study of three-dimensional turbulent channel flow at large reynolds numbers. J Fluid Mech 42:453–480

Devenport WJ, Simpson RL (1992) Flow past a wing-body junction—experimental evaluation of turbulence models. AIAA J. 30(4):873–881

Dol HS, Hanjalic K (2001) Computational study of turbulent natural convection in a side-heated near-cubic enclosure at a high rayleigh number. Int J Heat Mass Transf 44(12):2323–2244

Durbin PA (1991) Near-wall turbulence closure modeling without "Damping Functions". Theoret Comput Fluid Dyn 3(1):1–13

Durbin PA (1995) Separated flow computations with the K-E-V2 Model. AIAA J 33:659–664

Fleming JL, Simpson RL, Devenport WJ (1993) An experimental study of a turbulent wing-body junction and wake flow. Exp Fluids 14:366–378

Fu S, Rung T, Thiele F (1996) On the realizability of the nonlinear stress-strain relationship for reynolds-stress closures. Technical Report, HFI, TU-Berlin

Gatski TB, Speziale CG (1993) On explicit algebraic stress model for complex turbulent flows. J Fluid Mech 254:59–78

Germano M, Piomelli U, Moin P, Cabot WH (1991) A dynamic subgrid-scale eddy viscosity model. Phys Fluids A 3:1760–1765

Gibson MM, Launder BE (1978) Ground effects on pressure fluctuations in the atmospheric boundary layer. J Fluid Mech 86:491–511

Gibson MM, Rodi W (1981) A reynolds-stress closure model of turbulence applied to the calculation of a highly curved mixing layer. J Fluid Mech 103:161–182

Hah C, Lakshminarayana B (1980) Numerical analysis of turbulent wakes of turbomachinery rotor blades. J Fluids Eng 102:462–472

Hanjalic K, Kenjereš S, Durst F (1996) Natural convection in partitioned two dimensional enclosures at higher rayleigh numbers. Int J Heat Mass Transfer 39:1407–1427

Hossain MS, Rodi W (1982) A turbulence model for buoyant flows and its application to vertical buoyant jets. In turbulent buoyant jets and plumes

Huang PG, Bradshaw P, Coakley TJ (1992) Assessment of closure coefficients for compressible-flow turbulence models. NASA TM-103882

Jones WP, Launder BE (1973) The calculation of low-reynolds-number phenomena with a two-equation model of turbulence. Int J Heat Mass Transf 16:1119–1130

Jouvray A, Tucker PG (2005) Computation of the Flow in a ventilated room using non-Linear RANS, LES and hybrid RANS/LES. Int J Numer Meth Fluids 48(1):99–106

Jouvray A, Tucker PG, Liu Y (2007) On nonlinear RANS models when predicting more complex geometry room airflows. Int J Heat Fluid Flow 28:275–288

Keating A, Piomelli U (2006) A dynamic stochastic forcing method as a wall-layer model for large-eddy simulation. J Turbul 7(12)

Kolmogorov AN (1941) The local structure of turbulence in incompressible viscous fluid for very large reynolds number. Dokl Akad Nauk SSSR 30:299–303

Kolmogorov AN (1942) Equations of turbulent motion of an incompressible fluid. Izv Acad Sci, USSR, Phys 6(1–2):56–58

Lam CKG, Bremhorst K (1981) A modified form of the K-E model for predicting wall turbulence. J Fluids Eng 103:456–460

Launder BE (1989) Second-moment closure: present… and future? Inter. J Heat Fluid Flow 10(4):282–300

Launder BE, Sharma BI (1974) Application of the energy dissipation model of turbulence to the calculation of flow near a spinning disk. Letters in Heat Mass Transfer 1:131–138

Launder BE, Spalding DB (1974) The numerical computation of turbulent flows. Comput Methods Appl Mech Energy 3:269–289

Laurence DR, Uribe JC, Utyuzhnikov SV (2004) A robust formulation of the V2-F model. Flow Turbul Combust 73:169–185

Leschziner MA (1990) Modelling engineering flows with reynolds stress turbulence closure. J Wind Eng Ind Aerodyn 35:21–47

Li X, Yu Z, Zhao B, Li Y (2005) Numerical analysis of outdoor thermal environment around buildings. Build Environ 40:853–866

Lien FS, Durbin PA (1996) Non-linear K − E−V2 modeling with application to high-lift. Summer program proceedings. Center for Turbulence Research, NASA/Stanford Univ., 5–22

Lien FS, Kalitzin G (2001) Computations of transonic flow with the V2–F turbulence model. Int J Heat Fluid Flow 22:53–61

Lilly DK (1966) On the application of the eddy viscosity concept in the inertial sub range of turbulence. NCAR Manuscr, p 123

Lilly DK (1992) A proposed modification of the germano subgrid-scale closure model. Phys Fluids 4:633–635

Meneveau C, Katz J (2000) Scale-invariance and turbulence models for large-eddy simulation. Annu Rev Fluid Mach 32:1–32

Meneveau C, Lund TS, Cabot WH (1996) A lagrangian dynamic sub-grid scale model of turbulence. J Fluid Mech 319:353–385

Menter FR (1992) Improved two-equation K-W turbulence model for aerodynamic flows. ASA TM-103975

Menter FR (1994) Two-equation eddy-viscosity turbulence models for engineering applications. AIAA J 32:1598–1605

Morrison BI (2000) The adaptive coupling of heat and air flow modeling within dynamic whole-building simulation. Ph.D. Thesis, University of Strathclyde, Glasgow, UK

Moureh J, Flick D (2003) Wall air-jet characteristics and airflow patterns within a slot ventilated enclosure. Int J Therm Sci 42(7):703–711

Murakami S, Kato S, Kondo Y (1990) Examining K-E EVM by means of ASM for A 3-D horizontal buoyant jet in enclosed space. In: International Symposium on Engineering Turbulence Modelling and Measurements. ICHMT, Dubrovnik

Nielsen PV (1998) The selection of turbulence models for prediction of room airflow. ASHRAE Transactions, SF-98-10-1

Nieuwstadt FTM (1990) Direct and large-eddy simulation of free convection. In Proceeding of 9th International Heat Transfer Conference vol 1. Jerusalem, pp 37–47

Nieuwstadt FTM, Eggles JGM, Janssen RJA, Pourquie MBJM (1994) Direct and large-eddy simulations of turbulence in fluids. Futur Gener Comput Syst 10:189–205

Nilsson HO (2007) Thermal comfort evaluation with virtual manikin methods. Build Environ 42(12):4000–4005

Olcmen MS, Simpson RL (1995) An experimental study of a three-dimensional pressure-driven turbulent boundary layer. J Fluid Mech 290:225–262

Prandtl L (1925) Uber die ausgebildete turbulenz. ZAMM 5:136–139

Radmehr A, Patankar SV (2001) A new low-reynolds-number turbulence model for prediction of transition on gas turbine blades. Numer Heat Transfer, Part B 39:545–562

Renz U, Terhaag U (1990) Predictions of air flow pattern in a room ventilated by an air jet, the effect of turbulence model and wall function formulation. Proc. Roomvent '90: 18.1–18.15, Oslo

Reynolds O (1895) On the dynamical theory of incompressible viscous fluids and the determination of the criterion. Philos Trans R Soc Lond, A 186:123–164

Rodi W (1976) A new algebraic relation for calculating the reynolds stresses. ZAMM 56:219–221

Rodi W (1991) Experience with two-layer models combining the K-epsilon model with a one-equation model near the wall. AIAA, Aerospace Sciences Meeting, 29th, Reno, NV, Jan. 7–10, 1991. 13 P

Rodi W, Spalding DB (1984) A two-parameter model of turbulence and its application to separated and reattached flows. Numer Heat Transf 7:59–75

Roy C, Blottner F, Payne J (2003) Bluff-body flow simulations using hybrid RANS/LES. AIAA-2003–3889. In: 33rd AIAA fluid dynamics conference and exhibit, Orlando, Florida, June 23–26

Shih T, Liou W, Shabbir A, Yang Z, Zhu J (1995) A new K-E eddy viscosity model for high reynolds number turbulent flows. J Comput Fluids 24:227–238

Shur M, Spalart P, Strelets M, Travin A (1996) Navier–stokes simulation of shedding turbulent flow past a circular cylinder and a cylinder with backward splitter plate. In: Proceedings of the 3rd ECCOMAS computational fluid dynamics conference, 9–13 September, Paris, pp 676–682

Shur M, Spalart PR, Strelets M, Travin A (1999) Detached-eddy simulation of an airfoil at high angle of attack. In: 4th International Symposium on Engineering Turb. Modeling And Experiments, Corsica, France, May

Smagorinsky J (1963) General circulation experiments with the primitive equations I: the basic experiment. Month Wea Rev 91:99–164

Spalart PR, Allmaras S (1992) A one-equation turbulence model for aerodynamic flows. Technical Report AIAA-92-0439, AIAA

Spalart PR, Jou WH, Stretlets M, Allmaras SR (1997) Comments on the feasibility of LES for wings and on the hybrid RANS/LES approach. In: Proceedings of the first AFOSR international conference on DNS/LES

Spalding DB (1972) A two-equation model of turbulence. VDI-Forshungsheft 549:5–16

Speziale CG, Sarkar S, Gatski TB (1991) Modeling the pressure-strain correlation of turbulence: an invariant dynamical systems approach. J Fluid Mech 227:245

Squires KD (2004) Detached-eddy simulation: current status and perspectives. In: Proceedings of direct and large-eddy simulation-5

Srebric J, Chen Q, Glicksman LR (1999) Validation of a zero-equation turbulence model for complex indoor airflows. ASHRAE Trans 105(2):414–427

Tennekes H and Lumley JL (1972) A first course in turbulence. 1st edn. The MIT Press

Toraño J, Rodríguez R, Diego I (2006) Computational fluid dynamics (CFD) use in the simulation of the death end ventilation in tunnels and galleries. WIT Trans Eng Sci, 52

Tornstrom T, Moshfegh B (2006) RSM predictions of 3-D turbulent cold wall jets. Prog Comput Fluid Dyn 6:110–121

Travin A, Shur M, Strelets M, Spalart P (2000) Detached-eddy simulations past a circular cylinder. Flow Turbul Combust 63:293–313

Van Driest ER (1956) On turbulent flow near a wall. J Aeronaut Sci 23:1007–1011

Van Maele K, Merci B (2006) Application of two buoyancy-modified K–ε turbulence models to different types of buoyant plumes. Fire Saf J 41:122–138

Wilcox DC (1988) Reassessment of the scale-determining equation for advanced turbulence models. AIAA Journal 26:1299–1310

Wilcox DC (1994) Simulation of transition with a two-equation turbulence model. AIAA J 32:247–254

Wolfshtein M (1969) The velocity and temperature distribution in one dimensional flow with turbulence augmentation and pressure gradient. Int J Heat Mass Transfer 12:301–318

Xu W (1998) New turbulence models for indoor airflow simulation. Ph.D. Thesis, Department of Architecture, Massachusetts Institute Of Technology, Cambridge, MA

Yakhot V, Orszag SA (1986) Renormalization group analysis of turbulence. J Sci Comput 1:3–51

Zhai Z, Zhang Z, Zhang W, Chen Q (2007) Evaluation of various turbulence models in predicting airflow and turbulence in enclosed environments by CFD: Part-1: summary of prevalent turbulence models. HVAC&R Res 13(6):853–870

Zhang W, Chen Q (2000) Large eddy simulation of indoor airflow with a filtered dynamic subgrid scale model. Int J Heat Mass Transf 43(17):3219–3231

Zhang W, Chen Q (2005) Large Eddy simulation of the buoyancy flow driven by a corner heat source in a compartment. In: Proceedings of Indoor Air 2005, I(2): 1294–1299, Beijing, China

Zhang Z, Zhang W, Zhai Z, Chen Q (2007) Evaluation of various turbulence models in predicting airflow and turbulence in enclosed environments By CFD: Part-2: comparison with experimental data from literature. ASHRAE HVAC&R 13(6):871–886

Chapter 5
Select Numerical Methods

5.1 Discretization Methods (FDM, FVM, FEM)

The flow governing equations such as (3.47)–(3.50) with appropriate turbulence models, for instance, the zero-equation turbulence model (4.15) or the standard $k - \varepsilon$ model (4.18) and (4.19), as well as associated boundary conditions, create a complete system of flow equations. The flow governing equations need to be solved numerically due to their non-linear partial differential nature. The basic concept of the numerical simulation is to discretize the accurate, spatially and temporally continuous differential equations into approximate, discrete algebraic equations that can be solved by a computer. The discrete numerical values, instead of the continuous solutions, in the space concerned (computation domain), are then acquired.

Three primary numerical techniques widely used are: finite difference method (FDM) (Roache 1972), finite volume method (FVM) (Patankar 1980), and finite element method (FEM) (Baker 1983). The finite difference method discretizes the equations by expressing the derivatives with divided difference quotients. The finite volume method integrates conservation equations on each control-volume (cell) and replaces all cell fluxes with difference quotients. The finite element method employs so-called Galerkin weighted residual method to integrate the equations on each cell (Baker 1983; Patankar 1980).

The finite difference method is simple and easy to analyze the numerical accuracy of the discretization. But, it is generally difficult to interpret the physical meanings of some higher-order terms in the Taylor series. Meanwhile, the finite difference method does not guarantee the conservation of the discretized equations. In contrast, the finite volume method retains most physical information, and the conservation of the equations can be strictly achieved on every control-volume. The shortcoming of the finite volume method is that the accuracy of the discretization is difficult to analyze. The finite element method is powerful to handle complex geometries and is popular for solid mechanics analysis, but it is complicated to use due to its mathematical complexity (at least at the earlier time of CFD due to the computer limits). The method integrates the differential equations over an element or volume after having

© Springer Nature Singapore Pte Ltd. 2020
Z. Zhai, *Computational Fluid Dynamics for Built and Natural Environments*, https://doi.org/10.1007/978-981-32-9820-0_5

been multiplied by a weight function. The dependent variables are represented on the element by a shape function, which is the same form as the weight function. FEM's mathematical approach is generally difficult to put explicit physical significances on the terms in the algebraic equations, and thus requires special care to ensure a conservative flow solution. Although FEM is generally more stable than FVM, FEM requires more memory and has slower solution times than FVM.

Below is one example that is used to demonstrate the principles of these three methods: *to determine the horizonal temperature distribution in a vertically long flat plate with no internal heat source but given surface temperatures* as shown in Fig. 5.1. This is a 1-D steady-state heat conduction problem driven by surface temperature difference. Assume a constant thermal conductivity k.

(1) **Analytical solution**

$$\frac{d^2T}{dx^2} = 0 \tag{5.1}$$

$$T(x) = ax + b \tag{5.2}$$

Since T(x = 0) = 100 °C and T(x = 3) = 0 °C, a = −100/3 and b = 100. As a result: $T_1(x = 1/2) = 250/3 = 83.3333$; $T_2(x = 3/2) = 150/3 = 50$; $T_3(x = 5/2) = 50/3 = 16.6667$. It is a linear distribution of the inside temperatures.

(2) **FDM**

FDM uses the Taylor series to approximate various derivatives:

$$T(x + \Delta x) = T(x) + \Delta x\frac{dT}{dx} + \frac{1}{2}\Delta x^2\frac{d^2T}{dx^2} + \frac{1}{6}\Delta x^3\frac{d^3T}{dx^3} + o(\Delta x^4)\ldots \tag{5.3}$$

Fig. 5.1 One-dimensional heat conduction with given surface temperatures

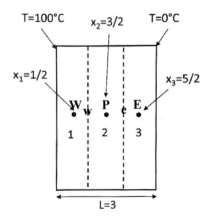

$$T(x - \Delta x) = T(x) - \Delta x\frac{dT}{dx} + \frac{1}{2}\Delta x^2\frac{d^2T}{dx^2} - \frac{1}{6}\Delta x^3\frac{d^3T}{dx^3} + o(\Delta x^4) \ldots \quad (5.4)$$

where Δx^4 and higher order terms are omitted when assuming Δx is infinitely (or adequately) small. Adding Eq. (5.3) with Eq. (5.4) provides:

$$T(x + \Delta x) + T(x - \Delta x) = 2T(x) + \Delta x^2\frac{d^2T}{dx^2} + o(\Delta x^4) \ldots \quad (5.5)$$

Therefore,

$$\frac{d^2T}{dx^2} = [T(x + \Delta x) + T(x - \Delta x) - 2T(x)]/\Delta x^2 + o(\Delta x^2) \ldots \quad (5.6)$$

Equation (5.1) can then be discretized as

$$T(x + \Delta x) + T(x - \Delta x) - 2T(x) = 0 \quad (5.7)$$

Equation (5.7) is the discrete format of the governing Eq. (5.1). The approximation using the Taylor series for the second derivative [Eq. (5.6)] has a truncation error term of the order Δx^2; hence this is considered second order accurate.

Equation (5.7) can be reorganized as

$$T(x) = \frac{T(x + \Delta x) + T(x - \Delta x)}{2} \quad (5.8)$$

In math, it represents a linear interpolation (average) of adjacent points for the middle value. Starting from initial guessed values for T_1, T_2, T_3, the inside temperature distribution can be iteratively updated with the given boundary conditions and the newest predictions. For a steady state problem, the initial values will not affect the final result but may influence the speed (i.e., iteration times) to reach a convergent result.

Equations (5.3)–(5.8) are yielded with a uniform grid Δx. For a general nonuniform grid such as the one in Fig. 5.1 that has $\Delta x_1 = 0.5$, $\Delta x_2 = 1$, $\Delta x_3 = 1$, $\Delta x_4 = 0.5$, the following general Taylor series is applied:

$$T(x_i + \Delta x_i) = T(x_i) + \Delta x_i\frac{dT}{dx} + \frac{1}{2}\Delta x_i^2\frac{d^2T}{dx^2} + \frac{1}{6}\Delta x_i^3\frac{d^3T}{dx^3} + o(\Delta x_i^4) \ldots \quad (5.9)$$

$$T(x_i - \Delta x_{i-1}) = T(x_i) - \Delta x_{i-1}\frac{dT}{dx} + \frac{1}{2}\Delta x_{i-1}^2\frac{d^2T}{dx^2} - \frac{1}{6}\Delta x_{i-1}^3\frac{d^3T}{dx^3} + o(\Delta x_{i-1}^4) \ldots$$
$$(5.10)$$

Dividing Eq. (5.9) and (5.10), respectively, by Δx_i and Δx_{i-1}, yields,

$$\frac{[T(x_i + \Delta x_i) - T(x_i)]}{\Delta x_i} = \frac{dT}{dx} + \frac{1}{2}\Delta x_i \frac{d^2T}{dx^2} + \frac{1}{6}\Delta x_i^2 \frac{d^3T}{dx^3} + o(\Delta x_i^3) \dots \quad (5.11)$$

$$\frac{[T(x_i - \Delta x_{i-1}) - T(x_i)]}{\Delta x_{i-1}} = -\frac{dT}{dx} + \frac{1}{2}\Delta x_{i-1} \frac{d^2T}{dx^2} - \frac{1}{6}\Delta x_{i-1}^2 \frac{d^3T}{dx^3} + o(\Delta x_{i-1}^3) \dots$$
$$(5.12)$$

Adding Eq. (5.11) with (5.12) and solving for $\frac{d^2T}{dx^2}$ provides:

$$\frac{d^2T}{dx^2} = \left\{ \frac{[T(x_i + \Delta x_i) - T(x_i)]}{\Delta x_i} + \frac{[T(x_i - \Delta x_{i-1}) - T(x_i)]}{\Delta x_{i-1}} + o(\Delta x_i^2) + o(\Delta x_{i-1}^2) \right\}$$
$$/ \frac{1}{2}(\Delta x_i + \Delta x_{i-1}) \quad (5.13)$$

Equation (5.13) has a truncation error term of the order Δx_i^2 and Δx_{i-1}^2. Equation (5.1) is then discretized as

$$\frac{[T(x_i + \Delta x_i) - T(x_i)]}{\Delta x_i} + \frac{[T(x_i - \Delta x_{i-1}) - T(x_i)]}{\Delta x_{i-1}} = 0 \quad (5.14)$$

$$T(x_i) = \frac{T(x_i - \Delta x_{i-1}) \times \Delta x_i + T(x_i + \Delta x_i) \times \Delta x_{i-1}}{(\Delta x_{i-1} + \Delta x_i)} \quad (5.15)$$

Equation (5.15) is the linear interpolation for a nonuniform grid. Table 5.1 shows two iteration processes with different initial values for the case in Fig. 5.1.

(3) **FVM**

FVM creates a volume around each node (1, 2 and 3 in Fig. 5.1) and integrates the differential equation over the volume:

$$\int \frac{d^2T}{dx^2}dv = \int \frac{d}{dx}\left(\frac{dT}{dx}\right)dv = 0 \quad (5.16)$$

Assume $dv = dx \times dA$ (dA is the face area of the volume),

$$\int d\left[\left(\frac{dT}{dx}\right)A\right] = 0 \quad (5.17)$$

$$\left(\frac{dT}{dx}\right)A\bigg|_e - \left(\frac{dT}{dx}\right)A\bigg|_w = 0 \quad (5.18)$$

e and w are the east and west face of each volume.
    To obtain the gradient at the faces, a linear interpolation is commonly used:

Table 5.1 Iterations of inside temperatures in heat conduction plate

T_{b1}	T_1	T_2	T_3	T_{b2}	T_{b1}	T_1	T_2	T_3	T_{b2}
100	0	0	0	0	100	50	50	50	0
100	66.66667	33.33333	11.11111	0	100	83.33333	66.66667	22.22222	0
100	77.77778	44.44444	14.81481	0	100	88.88889	55.55556	18.51852	0
100	81.48148	48.14815	16.04938	0	100	85.18519	51.85185	17.28395	0
100	82.71605	49.38272	16.46091	0	100	83.95062	50.61728	16.87243	0
100	83.12757	49.79424	16.59808	0	100	83.53909	50.20576	16.73525	0
100	83.26475	49.93141	16.6438	0	100	83.40192	50.06859	16.68953	0
100	83.31047	49.97714	16.65905	0	100	83.3562	50.02286	16.67429	0
100	83.32571	49.99238	16.66413	0	100	83.34095	50.00762	16.66921	0
100	83.33079	49.99746	16.66582	0	100	83.33587	50.00254	16.66751	0
100	83.33249	49.99915	16.66638	0	100	83.33418	50.00085	16.66695	0
100	83.33305	49.99972	16.66657	0	100	83.33362	50.00028	16.66676	0
100	83.33324	49.99991	16.66664	0	100	83.33343	50.00009	16.6667	0
100	83.3333	49.99997	16.66666	0	100	83.33336	50.00003	16.66668	0
100	83.33332	49.99999	16.66666	0	100	83.33334	50.00001	16.66667	0
100	83.33333	50	16.66667	0	100	83.33334	50	16.66667	0

$$\left(\frac{dT}{dx}\right)A\bigg|_e = \frac{T_E - T_P}{\Delta x_i} A_e \tag{5.19}$$

$$\left(\frac{dT}{dx}\right)A\bigg|_w = \frac{T_P - T_W}{\Delta x_{i-1}} A_w \tag{5.20}$$

where P is present node, E and W are adjacent East and West node, respectively. Taking Eqs. (5.19) and (5.20) into (5.18) yields:

$$\frac{T_E - T_P}{\Delta x_i} A_e - \frac{T_P - T_W}{\Delta x_{i-1}} A_w = 0 \tag{5.21}$$

For this 1-D case, assume $A_e = A_w$,

$$\frac{T_E - T_P}{\Delta x_i} - \frac{T_P - T_W}{\Delta x_{i-1}} = 0 \tag{5.22}$$

$$T_P = \frac{T_W \times \Delta x_i + T_E \times \Delta x_{i-1}}{\Delta x_i + \Delta x_{i-1}} \tag{5.23}$$

This is identical to Eq. (5.15). For a uniform grid with constant Δx_i for all i, Eq. (5.23) becomes

$$T_P = \frac{T_W + T_E}{2} \tag{5.24}$$

This is identical to Eq. (5.8). Note that Δx_i in FVM is the same as that in FDM, standing for the distance between two nodes (e.g., between 1 and 2 in Fig. 5.1).

Since the boundary conditions (surface temperature 100 and 0 °C) are on the faces of the volumes, instead of nodes, when calculating the values of nodes adjacent to the boundaries (i.e., 1 and 3 for the case in Fig. 5.1), node W is defined on face w for volume 1 and node E is on face e for volume 3 so as to take the impacts of the boundary conditions into the computational domain. Equation (5.23) should be used to account for the updated node distances at the boundaries. Similar iteration process as in FDM can be applied to obtain the same results.

(4) **FEM**

FEM starts with the energy conservation for each element (with two nodes: i and j). All the heats can only enter the system through the nodes (e.g., Q_i in Watts as shown in Fig. 5.2).

According to the Fourier law of heat conduction

$$q_i^{(e)} = -kA\frac{dT}{dx} = -kA\frac{T_j - T_i}{L^{(e)}} \tag{5.25}$$

From the energy conservation for the element (e)

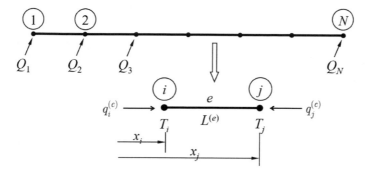

Fig. 5.2 Illustration of finite elements in heat transfer problem

$$q_j^{(e)} = -q_i^{(e)} = kA\frac{T_j - T_i}{L^{(e)}} \tag{5.26}$$

Combining Eqs. (5.25) and (5.26) provides

$$\begin{pmatrix} q_i^{(e)} \\ q_j^{(e)} \end{pmatrix} = \frac{kA}{L^{(e)}} \begin{pmatrix} 1 & -1 \\ -1 & 1 \end{pmatrix} \begin{pmatrix} T_i \\ T_j \end{pmatrix} \tag{5.27}$$

This is the element conduction equation. $\begin{pmatrix} 1 & -1 \\ -1 & 1 \end{pmatrix}$ is called the element conductance matrix.

For the case in Fig. 5.1, the following element conduction equations are created (T_{b1} and T_{b2} are respectively the boundary conditions of 100 and 0 °C):

$$\begin{pmatrix} q_{b1}^{(1)} \\ q_1^{(1)} \end{pmatrix} = \frac{kA}{L^{(1)}} \begin{pmatrix} 1 & -1 \\ -1 & 1 \end{pmatrix} \begin{pmatrix} T_{b1} \\ T_1 \end{pmatrix} \tag{5.28}$$

$$\begin{pmatrix} q_1^{(2)} \\ q_2^{(2)} \end{pmatrix} = \frac{kA}{L^{(2)}} \begin{pmatrix} 1 & -1 \\ -1 & 1 \end{pmatrix} \begin{pmatrix} T_1 \\ T_2 \end{pmatrix} \tag{5.29}$$

$$\begin{pmatrix} q_2^{(3)} \\ q_3^{(3)} \end{pmatrix} = \frac{kA}{L^{(3)}} \begin{pmatrix} 1 & -1 \\ -1 & 1 \end{pmatrix} \begin{pmatrix} T_2 \\ T_3 \end{pmatrix} \tag{5.30}$$

$$\begin{pmatrix} q_3^{(4)} \\ q_{b2}^{(4)} \end{pmatrix} = \frac{kA}{L^{(4)}} \begin{pmatrix} 1 & -1 \\ -1 & 1 \end{pmatrix} \begin{pmatrix} T_3 \\ T_{b2} \end{pmatrix} \tag{5.31}$$

The next step is to assemble all these element matrices into the global matrix using the heat conservation at nodes.

$$
\begin{pmatrix} Q_{b1} \\ Q_1 \\ Q_2 \\ Q_3 \\ Q_{b2} \end{pmatrix} = \begin{pmatrix} q_1^{(1)} \\ q_1^{(1)} + q_1^{(2)} \\ q_2^{(2)} + q_2^{(3)} \\ q_3^{(3)} + q_3^{(4)} \\ q_{b2}^{(4)} \end{pmatrix} = KA \begin{pmatrix} \frac{1}{L^{(1)}} & \frac{-1}{L^{(1)}} & 0 & 0 & 0 \\ \frac{-1}{L^{(1)}} & \frac{1}{L^{(1)}} + \frac{1}{L^{(2)}} & \frac{-1}{L^{(2)}} & 0 & 0 \\ 0 & \frac{-1}{L^{(2)}} & \frac{1}{L^{(2)}} + \frac{1}{L^{(3)}} & \frac{-1}{L^{(3)}} & 0 \\ 0 & 0 & \frac{-1}{L^{(3)}} & \frac{1}{L^{(3)}} + \frac{1}{L^{(4)}} & \frac{-1}{L^{(4)}} \\ 0 & 0 & 0 & \frac{-1}{L^{(4)}} & \frac{1}{L^{(4)}} \end{pmatrix} \begin{pmatrix} T_{b1} \\ T_1 \\ T_2 \\ T_3 \\ T_{b2} \end{pmatrix}
$$

$$(5.32)$$

For the case in Fig. 5.1, there is no internal heat source so $Q_1 = Q_2 = Q_3 = 0$; $T_{b1} = 100\ ^\circ\text{C}$ and $T_{b2} = 0\ ^\circ\text{C}$; Q_{b1} and Q_{b2} are unknown. $L^{(1)} = 0.5$, $L^{(2)} = 1$, $L^{(3)} = 1$, $L^{(4)} = 0.5$. Taking these into Eq. (5.32) yields

$$
\begin{pmatrix} Q_{b1} \\ 0 \\ 0 \\ 0 \\ Q_{b2} \end{pmatrix} = KA \begin{pmatrix} 2 & -2 & 0 & 0 & 0 \\ -2 & 3 & -1 & 0 & 0 \\ 0 & -1 & 2 & -1 & 0 \\ 0 & 0 & -1 & 3 & -2 \\ 0 & 0 & 0 & -2 & 2 \end{pmatrix} \begin{pmatrix} 100 \\ T_1 \\ T_2 \\ T_3 \\ 0 \end{pmatrix}
$$

$$(5.33)$$

Since KA are unknown and Q_{b1} and Q_{b2} are not asked, only the middle three equations need be solved:

$$
\begin{cases} -200 + 3T_1 - T_2 = 0 \\ -T_1 + 2T_2 - T_3 = 0 \\ -T_2 + 3T_3 = 0 \end{cases}
$$

$$(5.34)$$

Solving the Equation set (5.34) provides: $T_1 = 250/3$, $T_3 = 150/3$, $T_3 = 50/3$. The results are the same as the analytical ones as well as those from FDM and FVM.

The example above shows the general principles of each discretizing method, which produces the same results for the simple heat conduction case. The remaining of this chapter focuses on the introduction of finite volume method (FVM) that possesses both mathematical simplicity and physical conservation, and finite difference method (FDM) with the Taylor series theory that is effective in determining the accuracy of various numerical approximations.

5.2   Discretization of Governing Equations with FDM

The governing equations of incompressible turbulent flow (3.47)–(3.50) with the standard turbulence model Eqs. (4.18)–(4.19) can be generalized into the following uniform form:

$$
\frac{\partial \rho \phi}{\partial t} + \frac{\partial \rho U_j \phi}{\partial x_j} = \frac{\partial}{\partial x_j}\left(\Gamma_{\phi,\text{eff}} \frac{\partial \phi}{\partial x_j} \right) + S_\phi
$$

$$(5.35)$$

where ϕ represents the physical variable in question, as shown in Table 5.2. The equation has time, convection, diffusion and source terms.

The first step in the numerical procedure to solve the partial differential equations is to discretize the computational domain by dividing the whole flow field into a finite number of nodes or cells (control volume with the nodes at the center), as illustrated by a two-dimensional case in Fig. 5.3. This process is called "Grid Generation", which will be introduced further in Chap. 7. The continuous transport/conservation equations of mass, momentum and energy on the whole computational domain are then discretized to this discrete flow domain.

The Taylor series theory can be directly used to approximate various derivative terms in Eq. (5.35). The Taylor series for a general variable ϕ upon a general argument t has the following expressions:

$$\phi(t_i + \Delta t) = \phi(t_i) + \Delta t \frac{d\phi}{dt} + \frac{1}{2}\Delta t^2 \frac{d^2\phi}{dt^2} + \frac{1}{6}\Delta t^3 \frac{d^3\phi}{dt^3} + o(\Delta t^4) \qquad (5.36)$$

$$\phi(t_i - \Delta t) = \phi(t_i) - \Delta t \frac{d\phi}{dt} + \frac{1}{2}\Delta t^2 \frac{d^2\phi}{dt^2} - \frac{1}{6}\Delta t^3 \frac{d^3\rho}{dt^3} + o(\Delta t^4) \qquad (5.37)$$

The first order derivative $\frac{d\phi}{dt}$ can then be approximated with either a first order upwind differencing scheme (influenced by the upwind node or the previous time step) or a first order downwind differencing scheme (influenced by the downwind node or the next time step):

Table 5.2 Formula for the general form Eq. (5.35)

Equation	ϕ	$\Gamma_{\phi,\text{eff}}$	S_ϕ
Continuity	1	0	0
Momentum	U_i	$\mu + \mu_t$	$-\frac{\partial p}{\partial x_i} - \rho\beta(T - T_\infty)g_i$
Turbulent kinetic energy	k	$\mu + \frac{\mu_t}{\sigma_k}$	$\rho P + \rho G - \rho\varepsilon$
Dissipation rate of k	ε	$\mu + \frac{\mu_t}{\sigma_\varepsilon}$	$(C_{\varepsilon 1}\rho P + C_{\varepsilon 1}\rho G - C_{\varepsilon 2}\rho\varepsilon)\frac{\varepsilon}{k}$
Temperature	T	$\frac{\mu}{Pr} + \frac{\mu_t}{Pr_t}$	S_T
Concentration	C	$\frac{\mu}{Sc} + \frac{\mu_t}{Sc_t}$	S_C

Fig. 5.3 Two-dimensional illustration of computational domain discretization

 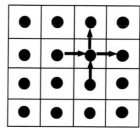

$$\frac{d\phi}{dt} = \frac{\phi(t_i) - \phi(t_i - \Delta t)}{\Delta t} + o(\Delta t) \tag{5.38}$$

$$\frac{d\phi}{dt} = \frac{(\phi t_i + \Delta t) - \phi(t_i)}{\Delta t} + o(\Delta t) \tag{5.39}$$

The first order derivative $\frac{d\phi}{dt}$ may also be approximated with a second order central differencing scheme that is obtained by subtracting Eq. (5.37) from (5.36):

$$\frac{d\phi}{dt} = \frac{\phi(t_i + \Delta t) - \phi(t_i - \Delta t)}{2\Delta t} + o(\Delta t^2) \tag{5.40}$$

Although having a higher order accuracy, the central differencing scheme is less used for approximating the first order derivative because of the inherent numerical instability.

The second order derivative $\frac{d^2\phi}{dt^2}$ is often approximated with a second order central differencing scheme that is obtained by adding Eq. (5.36) and (5.37):

$$\frac{d^2\phi}{dt^2} = \frac{\phi(t_i + \Delta t) - 2\phi(t_i) + \phi(t_i - \Delta t)}{\Delta t^2} + o(\Delta t^2) \tag{5.41}$$

Equations (5.38)–(5.41) can then be applied to discretize Eq. (5.35).

(1) Approximation of time derivative

In physics, flow at the current time step is influenced by the previous time step (history). According to Eq. (5.38), the first order approximation of $\frac{d(\rho\phi)}{dt}$ is:

$$\frac{d(\rho\phi)}{dt} = \frac{\rho\phi_P(t_i) - \rho\phi_P(t_i - \Delta t)}{\Delta t} = \frac{\rho\phi_P^n - \rho\phi_P^{n-1}}{\Delta t} \tag{5.42}$$

where ϕ_P stands for variable at present node, n is the current time step and $n - 1$ is the pervious time step.

(2) Approximation of convection term

Using Eq. (5.37), $\frac{\partial \rho U_j \phi}{\partial x_j}$ can be approximated with a simple upwind scheme:

$$\begin{aligned}
\frac{\partial \rho u_j \phi}{\partial x_j} &= \frac{\partial \rho u \phi}{\partial x} + \frac{\partial \rho v \phi}{\partial y} + \frac{\partial \rho w \phi}{\partial w} \\
&= \frac{(\rho u \phi)(x) - (\rho u \phi)(x - \Delta x)}{\Delta x} + \frac{(\rho v \phi)(y) - (\rho v \phi)(y - \Delta y)}{\Delta y} \\
&\quad + \frac{(\rho w \phi)(z) - (\rho w \phi)(z - \Delta z)}{\Delta z} \\
&= \frac{(\rho u \phi)_P - (\rho u \phi)_W}{\Delta x} + \frac{(\rho v \phi)_P - (\rho v \phi)_S}{\Delta y} + \frac{(\rho w \phi)_P - (\rho w \phi)_B}{\Delta z}
\end{aligned} \tag{5.43}$$

(3) **Approximation of diffusion term**

Equation (5.41) can be directly used to approximate the diffusion term $\frac{\partial}{\partial x_j}\left(\Gamma_{\phi,\text{eff}}\frac{\partial \phi}{\partial x_j}\right)$.

$$
\begin{aligned}
\frac{\partial}{\partial x_j}\left(\Gamma_{\phi,\text{eff}}\frac{\partial \phi}{\partial x_j}\right) &= \Gamma_{\phi,\text{eff}}\left(\frac{\partial^2 \phi}{\partial x^2} + \frac{\partial^2 \phi}{\partial y^2} + \frac{\partial^2 \phi}{\partial z^2}\right) \\
&= \Gamma_{\phi,\text{eff}}\left[\frac{\phi(x+\Delta x) - 2\phi(x) + \phi(x-\Delta x)}{\Delta x^2} + \frac{\phi(y+\Delta y) - 2\phi(y) + \phi(y-\Delta y)}{\Delta y^2}\right. \\
&\quad \left. + \frac{\phi(z+\Delta z) - 2\phi(z) + \phi(z-\Delta z)}{\Delta z^2}\right] \\
&= \Gamma_{\phi,\text{eff}}\left[\frac{\phi_E - 2\phi_P + \phi_W}{\Delta x^2} + \frac{\phi_N - 2\phi_P + \phi_S}{\Delta y^2} + \frac{\phi_T - 2\phi_P + \phi_B}{\Delta z^2}\right]
\end{aligned}
\tag{5.44}
$$

Substituting Eq. (5.42)–(5.44) into Eq. (5.35) yields the final form of discretized governing equations:

$$
A_P\phi_P = \sum_{nb} A_{nb}\phi_{NB} + S_U, \quad nb = w, e, s, n, b, t; \ NB = W, E, S, N, B, T \tag{5.45}
$$

$$
A_P = \sum_{nb} A_{nb} + \frac{\rho}{\Delta t} \tag{5.46}
$$

$$
S_U = S_\phi + \frac{\rho\phi_P^{n-1}}{\Delta t} \tag{5.47}
$$

Various direct and indirect numerical algebraic methods (as will be introduced in Chap. 8) can then be employed to solve this algebraic equation set.

5.3 Discretization of Governing Equations with FVM

FVM applies the conservation equations of fluid flow to each of discrete control volumes. Flow properties in each cell are assumed uniform. In practice, size of cells can be different (non-uniform grid) and geometry of cells can be nonorthogonal (body-fitted grid).

Integrating Eq. (5.35) over a typical control volume centered at P (Fig. 5.4 shows a two-dimensional projection of this cell on x-y plane, for clear demonstration) leads to a flux balance equation

$$
\int_{\Delta V} \frac{\partial(\rho\phi)}{\partial t}dV + I_e - I_w + I_n - I_s + I_t - I_b = \int_{\Delta V} S_\phi dV \tag{5.48}
$$

where I_f represents the total flux of ϕ across the cell-face f (= e, w, n, s, t, or b). Each of the surface fluxes I_f contains a convective contribution I_f^C and a diffusive contribution I_f^D, that is

Fig. 5.4 Two-dimensional
illustration of control
volumes and faces

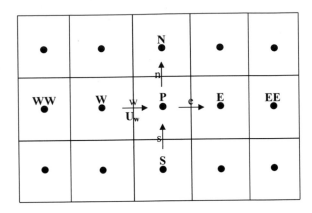

$$I_f = I_f^C + I_f^D \tag{5.49}$$

(1) Approximation of convection term

The convective contribution I_f^C in Eq. (5.49) can be approximated as:

$$I_f^C = C_f \phi_f \tag{5.50}$$

where C_f is the mass flux across the cell face f and can be calculated, for w-, s- and b-faces, as:

$$C_w = (\rho A u)_w, C_s = (\rho A v)_s, C_b = (\rho A w)_b \tag{5.51}$$

Below shows the process to obtain Eq. (5.50) and (5.51) by integrating the differential form of the convection term over the control volume $dv = dxdydz$:

$$
\begin{aligned}
\int \frac{\partial}{\partial x_j}(\rho U_j \phi) dv &= \int \frac{\partial}{\partial x}(\rho u \phi) dv + \int \frac{\partial}{\partial y}(\rho v \phi) dv + \int \frac{\partial}{\partial z}(\rho w \phi) dv \\
&= \int \partial[(\rho u \phi) dy dz] + \int \partial[(\rho v \phi) dx dz] + \int \partial[(\rho w \phi) dx dy] \\
&= [(\rho u \phi) dy dz]_e - [(\rho u \phi) dy dz]_w + [(\rho v \phi) dx dz]_n \\
&\quad - [(\rho v \phi) dx dz]_s + [(\rho w \phi) dx dy]_t - [(\rho w \phi) dx dy]_b \\
&= (\rho A u)_e \phi_e - (\rho A u)_w \phi_w + (\rho A v)_n \phi_n \\
&\quad - (\rho A v)_s \phi_s + (\rho A u)_t \phi_t - (\rho A u)_b \phi_b \\
&= C_e \phi_e - C_w \phi_w + C_n \phi_n - C_s \phi_s + C_t \phi_t - C_b \phi_b \tag{5.52}
\end{aligned}
$$

Proper determination of ϕ_f is essential for both accuracy and stability of numerical solutions. Different numerical schemes with different numerical accuracy orders are available to approximate ϕ_f at faces of each cell, such as, upwind differencing scheme, hybrid differencing scheme (Spalding 1972), QUICK differencing scheme (Leonard

1979), HLPA differencing scheme (Zhu 1991). Generally, more accurate schemes tend to be less stable, and vice versa. Taking the w-face of the control volume P as an example, the following briefly introduces the principles of several classical numerical schemes.

(a) **Upwind scheme** always approximates the variable value ϕ at each face (e.g., w) of a cell (e.g., P) according to its upwind cell value (Fig. 5.5), that is:

$$\phi_w = \begin{cases} \phi_w & \text{if } U_w > 0 \\ \phi_P & \text{if } U_w < 0 \end{cases} \tag{5.53}$$

Based on the Taylor series Eq. (5.3), this approximation has a truncation error term of the order Δx and hence has a first order accuracy.

(b) **Hybrid scheme** uses either central or upwind differencing (Fig. 5.5) according to a cell Peclet number $Pe = |C_w/D_w|$ (C_w is convective coefficient defined in Eq. (5.51) and D_w is diffusive coefficient defined in Eq. (5.67). Thus

$$\phi_w = \begin{cases} \frac{1}{2}(\phi_P + \phi_w) & \text{if } Pe \leq 2 \\ \phi_w & \text{if } Pe > 2 \end{cases} \tag{5.54}$$

Peclet number defines the ratio of convection to diffusion. When the convection dominates at a cell, the upwind node (depending on local flow direction) has a primary influence on the face value and thus the upwind scheme is appropriate. Otherwise, both the upwind and downwind nodes should be considered to calculate the face value. This ensures that the hybrid scheme has a second order accuracy.

Equation (5.5) can be rewritten as:

$$T(x) = \frac{[T(x + \Delta x) + T(x - \Delta x)]}{2} + \frac{1}{2}\Delta x^2 \frac{d^2T}{dx^2} + o(\Delta x^4) \dots \tag{5.55}$$

The is the origin of the central scheme that has a truncation error term of the order Δx^2.

Fig. 5.5 Illustration of using upwind and central scheme to calculate the face value ϕ_w ($\Delta x = 1$)

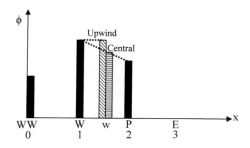

(c) **QUICK scheme** approximates the face value ϕ_w by fitting a parabolic curve through three nodal values ϕ_P, ϕ_w and ϕ_{ww}:

$$\phi_w = \frac{3}{8}\phi_P + \frac{3}{4}\phi_w - \frac{1}{8}\phi_{ww} \qquad (5.56)$$

Quick scheme stands for Quadratic Upwind Interpolation for Convective Kinetics and has a third order accuracy. Below present two approaches to obtaining the QUICK scheme, respectively, using the curve fitting technique and the Taylor series method.

- *Curve Fitting*

The face value ϕ_w can be approximated by fitting a parabolic curve through three nodal values ϕ_P, ϕ_w and ϕ_{ww} as shown in Fig. 5.6. Comparing this with the approximations in Fig. 5.5, the parabolic curve may better represent a nonlinear system. If Δx approaches infinite small, all numerical schemes should deliver the same results. Higher order scheme is preferred when a relatively coarse grid is used. This was why many efforts in developing robust high-order numeric schemes were found in 1970s–1990s when the computer power was weak so that coarse grid was a must.

Assuming $\phi = ax^2 + bx + c$ and applying this to WW, W and P yields

$$\phi_{ww} = a \times 0^2 + b \times 0 + c = c \qquad (5.57)$$

$$\phi_w = a \times 1^2 + b \times 1 + c = a + b + c \qquad (5.58)$$

$$\phi_P = a \times 2^2 + b \times 2 + c = 4a + 2b + c \qquad (5.59)$$

Solving for a, b and c provides:

$$\phi = ax^2 + bx + c = 0.5(\phi_P - 2\phi_w + \phi_{ww})x^2 + (2\phi_w - 0.5\phi_P - 1.5\phi_{ww})x + \phi_{ww} \qquad (5.60)$$

Substituting $x = 1.5$ gives the face value at w:

Fig. 5.6 Illustration of using parabolic curve fitting to calculate the face value ϕ_w ($\Delta x = 1$)

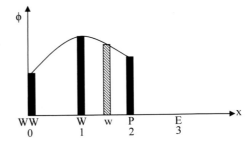

$$\phi_w = ax^2 + bx + c = 9/8(\phi_P - 2\phi_w + \phi_{ww})$$
$$+ 3/2(2\phi_w - 0.5\phi_P - 1.5\phi_{ww}) + \phi_{ww}$$
$$= 3/8\phi_P + 3/4\phi_w - 1/8\phi_{ww} \tag{5.61}$$

- **Taylor Series**

The Taylor series provides:

$$\phi(x+1) = \phi(x) + \frac{d\phi}{dx} + \frac{1}{2}\frac{d^2\phi}{dx^2} + \frac{1}{6}\frac{d^3\phi}{dx^3} + \cdots \tag{5.62}$$

$$\phi(x-1) = \phi(x) - \frac{d\phi}{dx} + \frac{1}{2}\frac{d^2\phi}{dx^2} - \frac{1}{6}\frac{d^3\phi}{dx^3} + \cdots \tag{5.63}$$

$$\phi(x-3) = \phi(x) - 3\frac{d\phi}{dx} + \frac{9}{2}\frac{d^2\phi}{dx^2} - \frac{27}{6}\frac{d^3\phi}{dx^3} + \cdots \tag{5.64}$$

If assume $\Delta x = 2$ in Fig. 5.6, and define $\phi(x) = \phi_w$ (face value), then $\phi(x + 1)$ $= \phi_P$, $\phi(x - 1) = \phi_w$, $\phi(x - 3) = \phi_{ww}$.

$$\frac{3}{8}\phi(x+1) + \frac{3}{4}\phi(x-1) - \frac{1}{8}\phi(x-3) = \frac{3}{8}\phi_P + \frac{3}{4}\phi_w - \frac{1}{8}\phi_{ww}$$
$$= \phi_w + 0 \times \frac{d\phi}{dx} + 0 \times \frac{d^2\phi}{dx^2} + \frac{36}{48}\frac{d^3\phi}{dx^3} + \cdots \tag{5.65}$$

The Taylor series method explicitly proves that the QUICK scheme takes the second order derivative into account but ignores the third order derivative; hence this is considered third order accurate.

(2) Approximation of diffusion term

Central differencing scheme is usually good for the approximation of second-derivative terms. By using central differencing scheme to the diffusion term in Eq. (5.49), I_f^D can be written, for the w-face as an example, as:

$$I_f^D = D_w(\phi_P - \phi_w) \tag{5.66}$$

where

$$D_w = \left(A^2\Gamma_\phi / \Delta V\right)_w \tag{5.67}$$

is the diffusive coefficient; A is w-face surface area and ΔV is cell volume.

Equation (5.66) is obtained by integrating the differential form of the diffusion term in Eq. (5.35) over the control volume $dv = dxdydz$:

$$\int \frac{\partial}{\partial x_j}\left(\Gamma_{\phi,\text{eff}}\frac{\partial}{\partial x_j}\right)dv = \int \frac{\partial}{\partial x}\left(\Gamma_{\phi,\text{eff}}\frac{\partial}{\partial x}\right)dv + \int \frac{\partial}{\partial y}\left(\Gamma_{\phi,\text{eff}}\frac{\partial}{\partial y}\right)dv$$

$$+ \int \frac{\partial}{\partial z}\left(\Gamma_{\phi,\text{eff}}\frac{\partial}{\partial z}\right)dv$$

$$= \int \partial\left[\left(\Gamma_{\phi,\text{eff}}\frac{\partial}{\partial x}\right)dydz\right] + \int \partial\left[\left(\Gamma_{\phi,\text{eff}}\frac{\partial}{\partial y}\right)dxdz\right]$$

$$+ \int \partial\left[\left(\Gamma_{\phi,\text{eff}}\frac{\partial}{\partial z}\right)dxdy\right]$$

$$= \left[\left(\Gamma_{\phi,\text{eff}}\frac{\partial}{\partial x}\right)dydz\right]_e - \left[\left(\Gamma_{\phi,\text{eff}}\frac{\partial}{\partial x}\right)dydz\right]_w$$

$$+ \left[\left(\Gamma_{\phi,\text{eff}}\frac{\partial}{\partial y}\right)dxdz\right]_n$$

$$- \left[\left(\Gamma_{\phi,\text{eff}}\frac{\partial}{\partial y}\right)dxdz\right]_s + \left[\left(\Gamma_{\phi,\text{eff}}\frac{\partial}{\partial z}\right)dxdy\right]_t$$

$$- \left[\left(\Gamma_{\phi,\text{eff}}\frac{\partial}{\partial z}\right)dxdy\right]_b \tag{5.68}$$

Subtracting Eq. (5.4) from Eq. (5.3) provides

$$T(x + \Delta x) - T(x - \Delta x) = 2\Delta x \frac{dT}{dx} + \frac{1}{3}\Delta x^3 \frac{d^3T}{dx^3} + o(\Delta x^4)\ldots \tag{5.69}$$

$$\left.\frac{dT}{dx}\right|_x = \frac{T(x + \Delta x) - T(x - \Delta x)}{2\Delta x} + o(\Delta x^2)\ldots \tag{5.70}$$

Equation (5.70) is a second order central differencing for the first order derivative. Applying Eq. (5.70) for all relevant terms in Eq. (5.68) yields:

$$\int \frac{\partial}{\partial x_j}\left(\Gamma_{\phi,\text{eff}}\frac{\partial}{\partial x_j}\right)dv = \left[\frac{(\Gamma_{\phi,eff}dydz)_e(\phi_E - \phi_P)}{dx_e}\right] - \left[\frac{(\Gamma_{\phi,eff}dydz)_w(\phi_P - \phi_W)}{dx_w}\right]$$

$$+ \left[\frac{(\Gamma_{\phi,eff}dxdz)_n(\phi_N - \phi_P)}{dy_n}\right]$$

$$- \left[\frac{(\Gamma_{\phi,eff}dxdz)_s(\phi_P - \phi_S)}{dy_s}\right] + \left[\frac{(\Gamma_{\phi,eff}dxdy)_t(\phi_T - \phi_P)}{dz_t}\right]$$

$$- \left[\frac{(\Gamma_{\phi,eff}dxdy)_b(\phi_P - \phi_B)}{dz_b}\right] = \left[\frac{(\Gamma_{\phi,eff}A^2)_e(\phi_E - \phi_P)}{dv_e}\right]$$

$$- \left[\frac{(\Gamma_{\phi,eff}A^2)_w(\phi_P - \phi_W)}{dv_w}\right] + \left[\frac{(\Gamma_{\phi,eff}A^2)_n(\phi_N - \phi_P)}{dv_n}\right]$$

$$- \left[\frac{\left(\Gamma_{\phi,eff} A^2\right)_s (\phi_P - \phi_S)}{dv_s} \right] + \left[\frac{\left(\Gamma_{\phi,eff} A^2\right)_t (\phi_T - \phi_P)}{dv_t} \right]$$

$$- \left[\frac{\left(\Gamma_{\phi,eff} A^2\right)_b (\phi_P - \phi_B)}{dv_b} \right]$$

$$= D_e(\phi_E - \phi_P) - D_w(\phi_P - \phi_W) + D_n(\phi_N - \phi_P) - D_s(\phi_P - \phi_S)$$

$$+ D_t(\phi_T - \phi_P) - D_b(\phi_P - \phi_B) \tag{5.71}$$

(3) Approximation of time derivative

The time derivative term in Eq. (5.35) can be approximated by different time discretization schemes. A simple first-order time scheme is

$$\frac{\partial(\rho\phi)}{\partial t} = \frac{(\rho\phi)^n - (\rho\phi)^{n-1}}{\Delta t} \tag{5.72}$$

where n refers to the current time step, $n - 1$ is the previous time step, and Δt is the time step increase. The volume integral of the time term can therefore be approximated as

$$\int_{\Delta V} \frac{\partial(\rho\phi)}{\partial t} dV = \frac{(\rho\phi)^n - (\rho\phi)^{n-1}}{\Delta t} \Delta V = \frac{\rho \Delta V}{\Delta t} \left(\phi_P^n - \phi_P^{n-1}\right) = S_P^T \phi_P^n - S_U^T \tag{5.73}$$

where $S_P^T = \frac{\rho \Delta V}{\Delta t}$, $S_U^T = \frac{\rho \Delta V}{\Delta t} \phi_P^{n-1}$.

A second order time scheme may also be considered, which however need store one more set of data from the previous time step $(n - 2)$.

$$\frac{\partial(\rho\phi)}{\partial t} = \frac{3(\rho\phi)^n - 4(\rho\phi)^{n-1} + (\rho\phi)^{n-2}}{2\Delta t} \tag{5.74}$$

(4) Approximation of source derivative

The source term S_ϕ is usually linearized as

$$S_\phi = S_\phi^U + S_\phi^P \phi_P \tag{5.75}$$

where, the coefficient S_ϕ^P is defined so that it is always not larger than zero for all the conservation equations. This operation enhances the stability of the numerical process (Patankar 1980). The volume integral of the source term can therefore be approximated as

$$\int_{\Delta V} S_\phi dV = S_\phi^U \Delta V + S_\phi^P \phi_P \Delta V = S_U' + S_P' \phi_P \tag{5.76}$$

where $S_U' = S_\phi^U \Delta V$, $S_P' = S_\phi^P \Delta V$.

(5) **Final form of discretized governing equations**

After replacing all the terms in Eq. (5.35) by their discretized analogues, the final form of the discretized governing equations results:

$$A_P \phi_P = \sum_{nb} A_{nb} \phi_{NB} + S_U, \text{ nb} = \text{w, e, s, n, b, t; NB} = \text{W, E, S, N, B, T} \quad (5.77)$$

where

$$A_P = \sum_{nb} A_{nb} - S_P \quad (5.78)$$

$$S_U = S'_U + S_U^T \quad (5.79)$$

$$S_P = S'_P - S_P^T \quad (5.80)$$

The main coefficients A_{nb} that relate the principal unknown ϕ_P to its immediate neighbors ϕ_{NB} contain the combined contribution from convection and diffusion. Equation (5.77) represents a set of algebraic equations describing the conservative features of the flow on each discrete cell. When higher order differencing schemes such as QUICK are used that consider the influences from farther neighboring nodes such as EE, WW, these terms are often grouped into the source term S_U so the general discretized Eq. (5.77) still stands.

In order to stabilize the solution process, it is usually necessary to under-relax the current solution by retaining part of the old solution:

$$\phi_P = \left(1 - \alpha_\phi\right)\phi_P^{old} + \alpha_\phi \phi_P \quad (5.81)$$

where $\alpha_\phi \in [0, 1]$ is an under-relaxation factor. Under-relaxation is a numerical practice that keeps part of the old value and only update the old value with a fraction (α) of the newly-obtained value. This practice can avoid a dramatic change between the old and new values (sometimes due to the numerical instability) that may lead to the divergence of the simulation. A smaller α results in a less update on the new value, which is good for physically or numerical instable cases; however may require more iterations to reach the final results. $\alpha = 0.5$ is often a good start considering the balance between computing efficiency and stability.

Introducing Eq. (5.81) into Eq. (5.77) leads to an under-relaxed difference equation that has the same form as Eq. (5.77) except that the coefficients A_P and S_U are replaced by:

$$S_U = S_U + \frac{1 - \alpha_\phi}{\alpha_\phi} A_P \phi_P^{old} \quad (5.82)$$

$$A_P = \frac{A_P}{\alpha_\phi} \quad (5.83)$$

5.4 Explicit and Implicit Method

Unsteady simulation includes a time term $\frac{\partial(\rho\phi)}{\partial t}$ that takes the historical impacts into the consideration. Two approaches are available to attain the current time step values from the historical values: the explicit and implicit methods.

(1) Explicit Method

The explicit method is to calculate the current P value ϕ_P^n with all historical values. The discretized governing equation in explicit format is expressed as:

$$\frac{(\rho\phi_P)^n - (\rho\phi_P)^{n-1}}{\Delta t}\Delta V + A_P\phi_P^{n-1} = \sum_{nb} A_{nb}\phi_{NB}^{n-1} + S \qquad (5.84)$$

$$\frac{(\rho\phi_P)^n}{\Delta t}\Delta V = -A_P\phi_P^{n-1} + \sum_{nb} A_{nb}\phi_{NB}^{n-1} + S + \frac{(\rho\phi_P)^{n-1}}{\Delta t}\Delta V \qquad (5.85)$$

Since all previous values (at $n - 1$) were obtained and stored already, the current value ϕ_P^n can be directly computed without iteration. However, the time step Δt used in the explicit method cannot be too large; in fact, it is constrained by the grid size Δx and local velocity U. According to the Courant–Friedrichs–Lewy (CFL) condition,

$$\Delta t < \frac{\Delta x}{U} \qquad (5.86)$$

The CFL condition is a necessary condition for convergence while solving certain partial differential equations in the explicit time integration scheme. This implies that the fluid movement in one time step cannot be more than one cell size. For a case with fine cells, this results in a much smaller time step (e.g., 0.001 s) than the actual time step of interest (e.g., 1 s). In another word, in order to obtain the result at $t = 1$ s, 1000 simulations on the time step $\Delta x = 0.001$ s are required. This computing cost may be comparable to (or even more than) the iteration effort required in the implicit method that allows the use of a large time step. Generally, if the transient details at a smaller time step are the goal of the simulation, the explicit method produce greater accuracy with less computational effort than the implicit method.

(1) Implicit Method

The implicit method is to calculate the current P value ϕ_P^n with all available newest values. The discretized governing equation in implicit format is expressed as:

$$\frac{(\rho\phi_P)^n - (\rho\phi_P)^{n-1}}{\Delta t}\Delta V + A_P\phi_P^n = \sum_{nb} A_{nb}\phi_{NB}^n + S \qquad (5.86)$$

$$\left(\frac{\rho^n\Delta V}{\Delta t} + A_P\right)\phi_P^n = \sum_{nb} A_{nb}\phi_{NB}^n + S + \left(\frac{\rho^{n-1}\Delta V}{\Delta t}\right)\phi_P^{n-1} \qquad (5.87)$$

Equation (5.87) is the same as Eq. (5.77). Equation (5.77) and those associated terms represent a fully implicit numerical solution process. Due to the coupling of ϕ_P^n with its neighbors ϕ_{NB}^n at the current time step (n), iteration is required to obtain the solution. The implicit method allows the use of large time steps and is more stable and robust than the explicit method, and thus is widely used in most CFD programs. Physically meaningful time step increase is still necessary to ensure an accurate and stable simulation even with the implicit method.

Practice-5: Indoor Airflow and Fire in a Long Tunnel

Example Project: Natural Ventilation in an Underground Shallow Traffic Tunnel

Background

Long and shallow underground vehicular tunnels are widely used in large cities to reduce surface traffic congestion. These tunnels often consist of two separate tubes, and each tube is unidirectional. These tubes are usually longer than 1000 m, with a height of approximately 4 m. Most of these tunnels are designed with series of roof openings for natural ventilation purpose, where toxic gases (e.g., polluted air or fire smoke) can be exhausted. Figure 5.7 shows one example of such a tunnel with roof openings (Tong et al. 2016). It is always questionable whether natural ventilation is adequate for ventilating indoor air and when mechanical ventilation is necessary.

Simulation Details

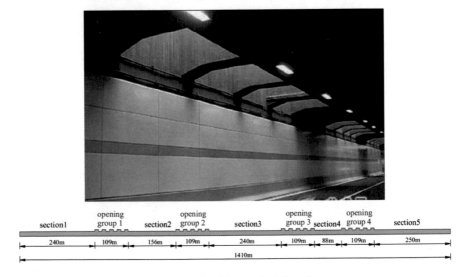

Fig. 5.7 One example of an underground traffic tunnel with roof openings

This study used both RANS and LES to simulate the airflow, heat transfer and fire smoke in a type tunnel as shown in Fig. 5.7. "Moving" cars were uniformly distributed on all three lanes in one tube and assumed to be traveling at same speed, which were simulated as "still" objects with momentum sources. Pressure boundary conditions were applied at both exits and all shaft openings. All solid surfaces were set to be no-slip and adiabatic conditions. Ambient wind was set at 0.95 m/s—the average of the field test. The standard k–ε turbulence model, and a second-order upwind discretization with unstructured grids were adopted. Figure 5.8 displays the computational domain and mesh. Fifteen multiple meshes were created including a fire domain (at a grid size of 0.083 m or 12 grids/m) and 14 non-fire domains (at a grid size of 0.167 m or 6 grids/m), and the total number of grid cells was 2,414,880. The solutions converged to a level at which the non-dimensional residuals of all equations were 10^{-4} or lower.

Results and Analysis

Figure 5.9 shows the predicted air velocity at one segment in the tunnel under the roof openings. It appears that the air velocities around the vehicles are higher, and the

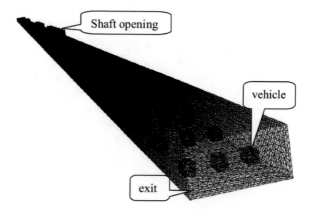

Fig. 5.8 Computational domain and mesh

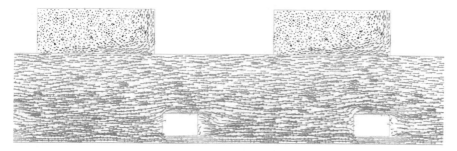

Fig. 5.9 Predicted air velocity at one segment in the tunnel under the roof openings

Table 5.3 Traffic conditions tested in experiment and CFD

Case	Vehicle spacing f (m)		Vehicle speed v_t (m/s)	
	Model experiment	Full-scale CFD	Model experiment	Full-scale CFD
1	0.8	8	1.6	5.1 (18 km/h)
2	0.8	8	2.0	6.3 (23 km/h)
3	1.3	–	1.6	–
4	1.3	13	2.0	6.3 (23 km/h)

airflows inside the openings are uneven and less evident but may still affect the traffic airflows. The study tested several traffic conditions as listed in Table 5.3 in experiment and CFD. Figure 5.10 compares the predicted mean air velocities at different tunnel locations with the experimental results. It is seen that all airflow directions are the same as the vehicle moving direction. Although in a similar trend, simulation results show some disparities from experiments, mostly due to the controllability of ambient wind in the model experiment. It is observed that the airflow velocity in the tunnel is weakened by the openings but is still adequate to ventilate the car emissions such as CO.

Two real fires were simulated with the total heat release rates (HRR) of 2.6 MW and 4 MW (the heat release rate per unit area of 1100 kW/m^2). Figure 5.11 shows the simulated results of smoke spreading for the 4 MW fire. The downstream smoke spreads faster than the upstream due to the influence of the ambient wind. After about 300 s, the upstream smoke does not move forward, and backflow occurs at the bottom of the front area. The smoke interface remains over the 1.8 m, which is comparable to the experimental result. Both experiment and simulation show that the smoke temperatures at downstream were higher than upstream but not at a life-risk level. Most of fire smoke can be exhausted from the roof openings. Ambient temperature has little influence on the smoke mass flow rates at the openings. Designing

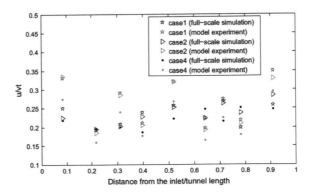

Fig. 5.10 Predicted and tested mean air velocities at different tunnel locations (normalized by vehicle speed v_t)

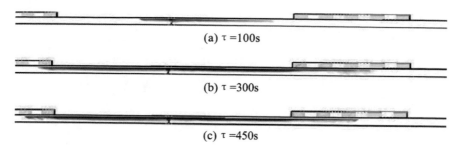

(a) τ =100s

(b) τ =300s

(c) τ =450s

Fig. 5.11 Predicted smoke spreading of the 4 MW fire

proper opening numbers per tunnel shaft is critical to ensuring sufficient and prompt ventilation.

Assignment-5: Simulating Natural Convection in a Confined Space

Objectives:

This assignment will use a computational fluid dynamics (CFD) program to model the temperature-difference-induced natural convection in a confined 2-D space.

Key learning point::

- Internal simulation
- Buoyancy effect
- Grid-independent solution.

Simulation Steps:

(1) Build a confined square space with L = 750 mm (as shown in Fig. 5.12);
(2) Prescribe boundary conditions for four walls (no-slip condition with standard wall function + given temperatures);
(3) Select a turbulence model: the standard k − ε model;
(4) Select a thermal effect model: the Boussinesq approximation, or, the ideal gas law
(5) Define convergence criterion: 0.1%;
(6) Set iteration: at least 2000 steps for steady simulation.

Cases to Be Simulated:

(1) Test four different grid resolutions (200 × 200, 300 × 300, 400 × 400, 500 × 500).

Report:

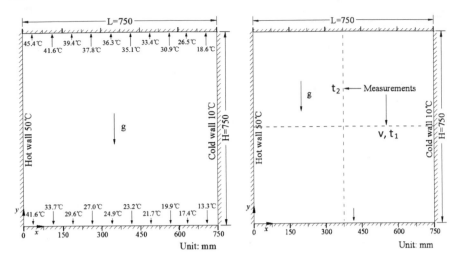

Fig. 5.12 Configuration, boundary condition setup and measurements of NC case

(1) Case descriptions: description of the cases
(2) Simulation details: computational domain, grid cells, convergence status

- Figure of the grids used (on X-Y plane);
- Figure of simulation convergence records.

(3) Result and analysis

- Figure of flow vectors;
- Figure of pressure contours;
- Figure of velocity contours;
- Figure of temperature contours;
- Evaluate the influences of grid resolution on simulation (using the process and index in Chap. 9);
- Validate the simulations with experimental data (Ampofo and Karayiannis, 2003).

(4) Conclusions (findings, result implications, CFD experience and lessons, etc.).

References

Ampofo F, Karayiannis TG (2003) Experimental benchmark data for turbulent natural convection in an air filled square cavity. Int J Heat Mass Transf 46(19):3551–3572
Baker AJ (1983) Finite element computational fluid mechanics. Hemisphere Publishing, New York
Leonard BP (1979) A stable and accurate convective modeling procedure based on quadratic interpolation. Comput Methods Appl Mech Eng 19:59–98
Patankar SV (1980) Numerical heat transfer and fluid flow. Hemisphere/McGraw-Hill, New York

Roache PJ (1972) Computational fluid mechanics. Hemosa, Albuquerque, New Mexico

Spalding DB (1972) A novel finite-difference formulation for differential expressions involving both first and second derivatives. Int J Numer Methods Eng 4:551–559

Tong Y, Zhai Z, Wang C, Zhou B, Niu X (2016) Possibility of using roof openings for natural ventilation in a shallow urban road tunnel. Tunn Undergr Space Technol 54:92–101

Zhu J (1991) A low diffusive and oscillation-free convection scheme. Commun Appl Numer Methods 7:225–232

Chapter 6
Specify Boundary Conditions

6.1 Classic Boundary Conditions

The governing equations of incompressible turbulent flow with the k-ε turbulence model can be expressed in the following general form:

$$\frac{\partial \rho \phi}{\partial t} + \frac{\partial \rho U_j \phi}{\partial x_j} = \frac{\partial}{\partial x_j}\left(\Gamma_{\phi,\text{eff}}\frac{\partial \phi}{\partial x_j}\right) + S_\phi \tag{6.1}$$

where ϕ represents the physical variable in question, as shown in Table 6.1. The equation has time, convection, diffusion and source terms.

To form a closed system of flow transport equations (ellipse equations) that can be solved mathematically, the boundary conditions at all the boundaries around the flow field are necessary. The variables to be solved in Eq. (6.1) (8 equations for 3-D flows) include: U_1, U_2, U_3, P, T, C, k, ε (8 variables). Due to the inherent relations between P and U_i (in the momentum equations), only three boundary conditions for P and U_i need be prescribed for 3-D flows (i.e., either 3 of U_i or P and 2 of U_i). All other boundary conditions for T, C, k, ε are required to mathematically enclose the associated governing equations. The accuracy of CFD prediction is highly sensitive to the boundary conditions supplied (assumed) by the user. The conventional flow boundary conditions for incompressible fluids include:

(1) *Inflow (inlet)*

At the inflow planes, the flow conditions, such as U, V, W, T, C, k, and ε, need to be specified based on the experiment or estimation. For example, for airflow inside buildings, the inflow boundary can be various types of air-supply diffusers with given flow velocity, temperature, contaminant concentration, and turbulences. Below is the general expression of inflow conditions:

$$\phi_{in} = \phi_{given} \tag{6.2}$$

© Springer Nature Singapore Pte Ltd. 2020
Z. Zhai, *Computational Fluid Dynamics for Built
and Natural Environments*, https://doi.org/10.1007/978-981-32-9820-0_6

Table 6.1 Formula for the general form Eq. (6.1)

Equation	ϕ	$\Gamma_{\phi,eff}$	S_ϕ
Continuity	1	0	0
Momentum	U_i	$\mu + \mu_t$	$-\frac{\partial p}{\partial x_i} - \rho\beta(T - T_\infty)g_i$
Turbulent kinetic energy	k	$\mu + \frac{\mu_t}{\sigma_k}$	$\rho P + \rho G - \rho\varepsilon$
Dissipation rate of k	ε	$\mu + \frac{\mu_t}{\sigma_\varepsilon}$	$(C_{\varepsilon 1}\rho P + C_{\varepsilon 1}\rho G - C_{\varepsilon 2}\rho\varepsilon)\frac{\varepsilon}{k}$
Temperature	T	$\frac{\mu}{Pr} + \frac{\mu_t}{Pr_t}$	S_T
Concentration	C	$\frac{\mu}{Sc} + \frac{\mu_t}{Sc_t}$	S_C

where ϕ stands for various variables in question. Note that negative flow velocity is allowed, which represents an outflow condition in physics. The turbulence properties should also be provided, which can be measured or estimated from experience. The following estimations are commonly used for k and ε:

$$k_{in} = C \times U_{in}^2 \tag{6.3}$$

$$\varepsilon_{in} = k_{in}^{3/2}/(\alpha H) \tag{6.4}$$

where C is a constant (5–20%) depending on turbulence intensity of inflow; α is an empirical coefficient (usually 0.2) and H is the characteristic length of inlet (e.g., diffuser opening size).

(2) *Outflow (outlet)*

The outflow conditions can also be explicitly specified if they are known, such as the exhaust outlet powered by a mechanical fan, using the above inflow setting method. If the outflow conditions are undetermined, at the outflow planes, the streamline gradients of all variables can be set to zero, implying a fully developed flow condition.

$$\frac{\partial \phi}{\partial n} = 0 \tag{6.5}$$

where ϕ stands for U, V, W, T, C, k, or ε; and n is the coordinate normal to the outflow plane. Obviously, the fully developed flow condition may not be suitable for some outflow boundaries, for instance, windows in a building, where the airflow conditions vary with locations. In these locations, either inflow or pressure boundary condition (to be introduced later) should be specified and used.

(3) *Symmetry surface*

The symmetry boundary condition is useful to reduce the computation domain and the computing effort for the flows that have symmetrical geometries, boundary conditions and flow patterns. Since there is no flow across the symmetry plane (no-penetration condition) and all variables are symmetric at both sides of the plane, the boundary condition can be expressed as

$$V_n = 0 \tag{6.6}$$

and

$$\frac{\partial \phi}{\partial n} = 0 \tag{6.7}$$

where V_n is the normal velocity component to the symmetry surface; ϕ are the variables other than V_n and n is the coordinate normal to the symmetry surface.

(4) *Rigid surface*

Rigid surfaces do not allow the penetration of flow through the surfaces. Such examples include ground, wall, ceiling, floor, and surfaces of furniture, appliances, and occupants. Rigid surface can be either stationary or moving. With the no-slip condition,

$$\vec{V}_{fluid\,at\,rigid} = \vec{V}_{rigid} \tag{6.8}$$

$\vec{V}_{rigid} = 0$ for stationary rigid surfaces. Three types of thermal boundary conditions can be provided for rigid surfaces:

(a) Dirichlet condition: $T = T_{rigid}$ (known rigid surface temperature)
(b) Neumann condition: $\partial T / \partial n = \dot{q} / k_{fluid}$ (known heat flux $\dot{q}$ at rigid surface—W/m^2)
(c) Robbins condition: $\partial T / \partial n = h / k_{fluid} \times (T_{fluid} - T)$.

Here, k_{fluid} is the thermal conductivity of fluid; h is the convective heat transfer coefficient; T_{fluid} is the temperature of fluid away from the thermal boundary layer of the rigid surface. The same three types of boundary conditions can be prescribed for the concentration equation, although the non-penetration condition is mostly used: $\partial C / \partial n = 0$ (Neumann condition) that implies a bounce-back model of species at a rigid surface.

For the turbulent flow around rigid surfaces, besides the common no-slip condition for viscous flows, special near-wall treatment techniques need be used to describe the low-Reynolds turbulent flow and heat transfer at this region. Those methods include, as discussed before in Chap. 4, the wall function laws (e.g., Launder and Spalding 1974), two-layer models (e.g., Rodi 1991), the low-Reynolds-number models (e.g., Launder and Sharma 1974).

The following is a typical expression of wall function, which considers the near-wall flow as laminar if $y^+ \leq 11.63$ and as turbulent if $y^+ > 11.63$, as illustrated in Fig. 6.1. In physics, when the flow is very close to the surface/wall (i.e., in the viscous sub-layer of the boundary layer), the flow is slow and laminar and thus a linear velocity profile can be applied (as shown by the sloped dash line in Fig. 6.1). When the flow moves slightly away from the surface ($y^+ > 11.63$), a log law profile of velocity is observed (attributed to the near wall turbulence impacts) and thus imposed

Fig. 6.1 Velocity
distribution near a rigid
surface

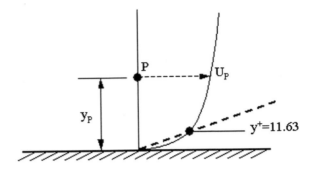

for the computational cells (nodes) in this region. The wall function laws for velocity
and temperature can be generally written as:

$$u^+ = y^+ \qquad \text{for } y^+ \leq 11.63 \tag{6.9}$$
$$T^+ = Pr\, y^+$$

$$u^+ = \tfrac{1}{\kappa} \ln\!\left(E\, y^+\right)$$
$$T^+ = Pr_t\!\left[u^+ + E\!\left(\tfrac{Pr}{Pr_t} - 1\right)\!\left(\tfrac{Pr_t}{Pr}\right)^{1/4}\right] \qquad \text{for } y^+ > 11.63 \tag{6.10}$$

and

$$u^+ = \frac{U_P}{u_\tau},\ u_\tau = \sqrt{\frac{\tau_w}{\rho}},\ y^+ = \frac{y_P u_\tau \rho}{\mu},\ T^+ = \frac{(T_w - T_P)\, u_\tau}{q} \tag{6.11}$$

where u^+ is the dimensionless mean velocity, U_P is the velocity parallel to the wall at
point P, u_τ is the friction velocity, τ_w is the wall shear stress, y^+ is the dimensionless
wall distance, y_P is the distance to the rigid surface, κ is the von Karman constant
($\kappa = 0.41$), E is the roughness parameter (E = 9 for hydraulically smooth walls
with constant shear stress), T_w is the rigid surface (wall) temperature, T_P is the air
temperature at point P, and q is the heat transfer rate (heat flux).

The development of wall functions adopts many assumptions, such as, Prandtl
mixing hypothesis, Boussinesq eddy-viscosity assumption, fully developed flow,
and no pressure gradients or other momentum sources (constant shear stresses). For
airflows in confined spaces, most of these assumptions are acceptable; therefore, wall
functions are still widely used. However, the heat transfer in the near-wall region may
not be accurately predicted by this approach. Improved near-wall turbulence models
may be required to obtain the correct heat transfer from the wall boundaries (Chen
1988).

(1) *Periodic Boundary*

Periodic boundary condition can be useful to model a flow with periodic conditions
(e.g., the outflow is fed back to the inflow) in terms of saving computational efforts.
One classic example is to model the flow in an infinitely long pipe/duct. Instead of

Fig. 6.2 Illustration of
pressure boundary setting
and calculation

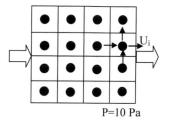

P=10 Pa

simulating such an infinitely (or extremely) long pipe/duct flow, a pipe/duct section
is often modeled with inflow conditions set to be the same as the calculated outflow
conditions.

$$\phi_{in} = \phi_{out} \qquad (6.12)$$

where ϕ stands for various variables. Equation (6.12) works for general periodic
boundaries.

(1) *Pressure Boundary*

Pressure such as atmosphere pressure at boundaries, if given, can also be prescribed.

$$P = P_{given} \qquad (6.13)$$

This fixed pressure will be used in the momentum equations to solve the velocity
field. Note that only relative pressure difference is important to drive the fluid flow and
absolute pressure is not necessary. The prediction on absolute pressure distributions,
if needed, will require a reference pressure (e.g., standard atmosphere pressure atm)
that will be added to all calculated pressure values in the domain. Once pressure is
provided at a boundary, only two of the three velocity components (U, V, W) are
required to close the equations, the other one will be calculated to ensure the mass
conservation at the boundary, as illustrated below (Fig. 6.2) where $P = 10$ Pa is set at
the exit column with the assumption of $V = W = 0$. U_i at the exit will be computed
for each cell i using the mass conservation equation. In another word, the enforced
pressure does not guarantee that the momentum equation yielded velocities at the
boundary meet the mass conversation.

6.2 Practical Boundary Conditions

Boundary conditions are crucial for the accuracy of the CFD prediction. The boundary conditions specified in CFD can be obtained from measurements. But in practice
most of them are based on empirical data or even experienced assumption. For
instance, for a building design project, the interior surface temperatures for the walls
and roofs are often estimated, and sometimes may even use the adiabatic assumption

for simplification. The circumstance may become worse when time-varying boundary conditions are needed for an unsteady calculation, in which the dynamic measure data is usually unavailable and even the estimate is difficult to make.

(1) *Wind Profile*

Because of the no-slip condition of the ground, the wind speed illustrates a gradual change from zero to ambient speed depending on terrain status. Equation (6.14) is a commonly used wind profile:

$$U_{wind} = U_{ref} \left(\frac{Z}{Z_{ref}} \right)^n \tag{6.14}$$

where U_{wind} is the wind velocity at height Z; U_{ref} is the reference wind speed at a reference height Z_{ref} (typically 10 or 40 m) (i.e., ambient wind speed). n is the wind profile index depending on terrain conditions (e.g., 0.14 for plat field and 0.28 for city center). If U_{wind} is parallel to U in the computational domain, $V = W=0$ at the inlet boundary. Otherwise, the setting of U, V, W at inlet should consider the wind direction (i.e., angles to the flow domain).

(2) *Diffusers*

Srebric and Chen (2002) indicated that it is not possible to use standard jet formulae to provide boundary conditions of various air supply diffusers for modeling airflows in confined spaces. Because the effective supply area A_{eff} (i.e., supplying openings/holes) is less than the gross diffuser area A_d, the resultant air momentum (i.e., velocity) with a given mass flow rate is strong in reality. The momentum method proposed by Chen and Moser (1991) is often employed to de-couple the momentum and mass boundary conditions for the diffuser in a CFD simulation. The diffuser is represented in CFD as an opening that has the same gross area, mass inflow rate, and momentum flux as a real diffuser. This model enables specification of the source terms (fixed values) in the conservation equations over the gross diffuser area. The supply velocity for the momentum source term is calculated from the mass flow rate, $\dot{m}$, and the diffuser effective area A_{eff}:

$$V_{supply} = \dot{m}/(\rho A_{eff}) \tag{6.15}$$

The momentum method requires the following data as the boundary conditions:

- flow rate
- discharge jet velocity or effective diffuser area
- supply turbulence properties
- supply temperature and contaminate concentrations.

If a diffuser has a specific supply angle (or angles) (as showed in Fig. 6.3), specifying velocities at individual cells becomes inevitable. Either average or area-weighted velocity could be used. The grid cell spatial resolution depends on the trade-off between simulation accuracy requirement and input effort. Most jet flows have direct

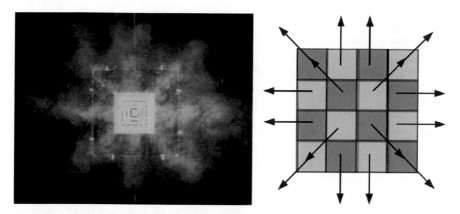

Fig. 6.3 Flow visualization of square ceiling diffuser (left) and velocity angles for the momentum method (right) (Srebric and Chen 2002)

impacts on the adjacent flow field (~1–3 m). If the space is significantly large (e.g., over 10 m) and the study focus is on the main flow field (rather than the jet area), the influence of jet model may be less critical. Otherwise, an accurate jet/diffuser model is always desirable. Srebric and Chen (2002) presented and discussed proper simplified numerical models for complex air supply diffusers.

(3) *Isothermal and Thermal Objects*

Solid objects in the flow domain such as building, vehicle, furniture, human, appliances are often modeled as blockages, which bring not only resistance to the flow but also mass and heat transfer. These blockages can be included in the CFD calculation, which is called conjugate heat transfer problem – that predicts variations of temperature within solids and fluids due to the interaction between solids and fluids. The governing equations for fluids still work for solids except that all velocity-related terms disappear due to the no-movement state inside static solids.

For isothermal cases, blockages mostly serve as the resistance to the flow and thus the surface roughness may be needed. For non-isothermal cases, the heat conduction in solid can be calculated, which requires the specifications of solid properties such as density, heat capacity, thermal conductivity etc. The heat transfer between the solid and fluid is through the convection at the external surface of the solid. The convective heat transfer coefficient can be computed by CFD, which however is highly sensitive to the grid quality and turbulence model used (Zhai and Chen 2004). An alternative is to provide a user-set convective heat transfer coefficient. A constant body temperature or total heat flux (in Watt) or unit heat flux (in Watt/m^2) can be prescribed for the solid, which will then interfere with the surrounding fluid.

To simply the calculation and accelerate the convergence, blockages can also be excluded from the CFD simulation. The influences of blockages to the fluid are primarily via the surfaces of the solids. As a result, the boundary condition treatment method for rigid surfaces can be applied. A body is simply a closed connection of rigid surfaces. The grid cells within each body have a speed of zero or share the body speed (if the solid is moving) by default. Either actual or simplified geometries of objects can be utilized, depending on the availability of the geometries (e.g., CAD file), simulation requirements, and computing effort expectation etc. In general, more accurate the model, more accurate the simulation, and more effort the modeling.

Practice-6: Simulation of Computer Rack in Data Center

Example Project: Simplified Rack Boundary Conditions for Data Center Models
Background:
Data centers are energy suckers due to high electricity demand for both intensive computing and cooling. The layout and design of a data center can make a significant difference in its energy use and the consequences of improper data center design can be dramatic. Cooling energy in poorly designed data centers can constitute up to 50% of its energy use. CFD plays an important role in aiding the layout design and management of data centers. While the use of CFD modeling is common in data center design, there are some important issues that need to be resolved.

Modeling the computer/server rack is one of the critical pieces in the design process. Often this is done as a black box rather than modeling the rack in detail. Modeling a computer rack as a black box has been carried out in numerous data center studies, but rarely has it been validated against experimental temperature and velocity data. In a black box model, room conditions are put into the front (inlet) of the rack and the added enthalpy outputs come from the back of the rack (outlet). One of the central issues with this approach is the question of which boundary conditions for the rack produce acceptable accuracy. The goal of this project was to develop a set of easily reproducible boundary conditions and validate them against sets of experimental data.

Simulation Details:
His study developed two distinct CFD models, an open box model (OBM) and a black box model (BBM). Both models were designed to be simple and require minimal user inputs. All models were developed using commercially available CFD software. The OBM was developed first and its purpose was to be an interim step to inform the development of the BBM. While it was very simplified in its detail, the OBM was still an approximation of the server simulator and allowed for air to flow through the rack model. The BBM, by contrast, was a solid box. It took inputs at the rack inlet and outputted modified values at the rack outlet. Its assumptions were tested both against the experimental data and the OBM. Both models shared the Boussinesq approximation and the κ-ε turbulence model used in the RANS equations. All surfaces were modeled as adiabatic surfaces and radiation models were not used.

(1) *Open Box Model (OBM)*

Figure 6.4 shows the layout of the experiment setup and the Open Box Model (OBM). All sides of the rack were modeled as adiabatic plates. The front and back plates were modeled with a percent open area that allowed restricted airflow to pass through. Both the fan and heating plates were broken up to allow different heat fluxes and flow rates for each server simulator. The fan plates were given defined X velocities and were modeled without any swirl. The heating plates were given a defined heat flux (divided

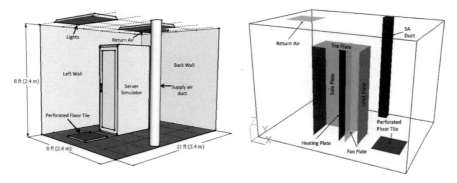

Fig. 6.4 Layout of experiment setup and open box model (OBM)

among each server simulator) with one-half of the heat flux coming out of each side of the plate.

(2) *Black Box Model (BBM)*

The rack inlet boundary conditions were defined by a velocity normal to the front rack plate (i.e., the rack door) face and a temperature profile. A uniform velocity was imposed normal to the plate. Smoke pen tests conducted during the experiment as well as validated models confirmed this as a reasonable assumption. At the rack inlet face, this velocity was defined by the fan speed divided by the porosity as shown in Eq. (6.16). The same velocity condition was imposed at the rack exhaust boundary plate.

$$V_{face} = \frac{V_{fan}}{plate\ porosity} \tag{6.16}$$

where V_{face} is velocity normal to the rack inlet or exhaust at the plate (m/s); V_{fan} is velocity imposed by the rack fans (m/s); plate porosity is expressed as a percent open area of the rack door (%).

At the rack inlet plate, the temperature was taken from the adjacent upwind cell. For the rack exhaust plate, this temperate (at the same y and z coordinates) was taken with the appropriate amount on enthalpy added, shown in Eq. (6.17).

$$T_{ex} = T_{in} + \frac{q_{server}}{c_p \cdot \dot{m}} \tag{6.17}$$

where T_{ex} is exhaust temperature for cell on rack exhaust plate (°C); T_{in} is temperature at rack inlet plate for same y, z (°C); q_{server} is heat added by server (W) (for BBM total heat generated by each server was assumed to be evenly distributed over the server cross sectional area); c_p is specific heat capacity of air (J/(kg-°K)); $\dot{m}$ is mass flow rate across the cell (kg/s).

Figure 6.5a shows the set-up for the BBM while Fig. 6.5b shows the translation of temperatures and velocity values from the front plate to the rear plate.

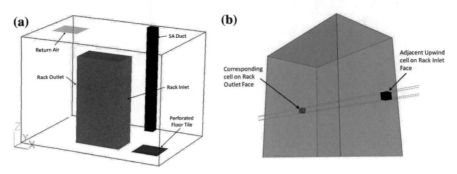

Fig. 6.5 Black box model (BBM)

Results and Analysis:

(1) *Grid Independence*

The normalized root mean squared error (NRMSE) was used to analyze the results of different levels of meshes to find the grid independent solution. Resolutions of 72,000, 244,800, 576,000, and 1,150,000 cells were tried. Grid independence was found at 244,800 cells with the average cell length coming out to approximately 4 cm (as illustrated in Fig. 6.6).

(2) *Temperature and Velocity Agreement*

Both OBM and BBM models were tested across a range of airflows and rack loads. Overall, good agreements were observed between the modeling and the test. The average temperature agreement between all experiments and model results were within 2.9 °C, on average. Velocity predictions across all models were found to be within 0.2 m/s, on average. There was very little difference between the results for the open box model and the black box model. This was considered to be a good sign since it indicates that black box programming is not necessarily required to produce good modeling results. The only caveat of using an open box model is that people who read the results need to understand that while it allows airflow through the rack

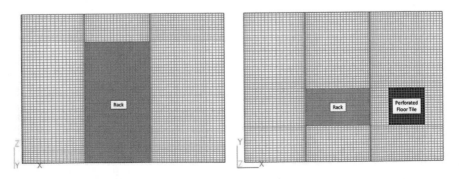

Fig. 6.6 Y-Plane mesh at middle of rack (left) and Z-Plane mesh at middle height of rack (right)

for purposes of the simulation, it is not intended to give results for rack-internal airflow and heat transfer—only room level results.

One select comparison of simulation with experiment is presented below to give a representative sample of the full range of rack experiments. Temperature results shown below are normalized against the supply air temperature and the exhaust air temperature. While temperature rise across a rack is a more familiar metric for those in the server industry, the study chose to use the supply air temperature instead of the rack inlet temperature as the lower boundary. This is due to the fact that the supply air will always be the lowest temperature in the room. Just looking at the temperature rise across the rack does not let one compare two equal racks where the supply air temperature might be different. The temperature rise across the racks should be the same across both racks, but room conditions could be significantly different as a result of the different supply air temperatures. Therefore, to make comparisons more universal, this study wanted to consider the more encompassing temperature boundaries for the room. Velocity results were examined in an absolute sense. This was due to the fact that velocity was a much more difficult and uncertain measurement to take and therefore was considered more of a secondary comparison.

Experiment 1: 4 kW Rack, Fan Speed of 0.56 m/s
The 4 kW Rack had an even power distribution and a fan speed setting of 0.56 m/s. Figure 6.7 displays the predicted temperature and velocity by both OBM and BBM.

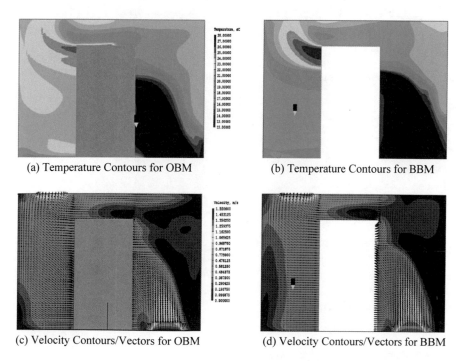

(a) Temperature Contours for OBM   (b) Temperature Contours for BBM

(c) Velocity Contours/Vectors for OBM   (d) Velocity Contours/Vectors for BBM

Fig. 6.7 Predicted temperature and velocity contours/vectors for experiment 1

Cool air supplied from the perforated floor title is induced into the rack front due to
the interior server fans and exists through the rack back with a higher temperature
at the upper of the rack. Figure 6.8 shows normalized temperature results for the
vertical poles at the immediate front and back of the rack. Under-prediction was
noticed for both poles. For the front pole, this under-prediction may be due to a
slight under-prediction of the throw for the perforated floor tile. The effects of the
under-predictions of the models for the rack-inlet pole may have carried over to some
of the under-predictions on the rack outlet pole. Figure 6.9 shows the velocity pole
comparisons for this experiment. The disparity between simulation and experiment

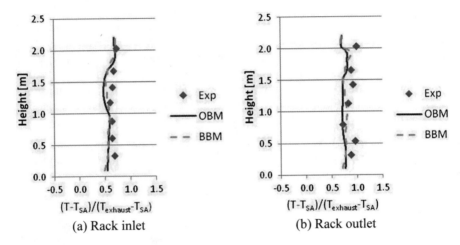

Fig. 6.8 Comparison of air temperature at the immediate front and back of the rack

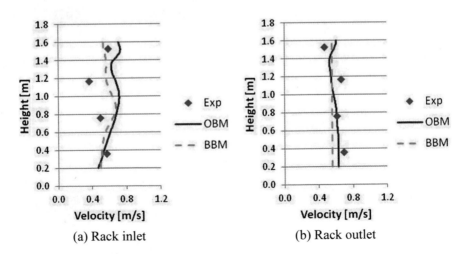

Fig. 6.9 Comparison of air velocity at the immediate front and back of the rack

is mostly attributed to the challenge in modeling the supply throw from perforated floor tile.

In general, both the open box model and black box model produce acceptable results as validated against ten different sets of experimental data for a rack populated by four 10 U server simulators. The steps for setting up the boundary conditions for a black box rack model from this study are easily reproducible and require minimal user inputs of rack load and airflow. It is hoped that these steps will give data centers designers a better ability to develop models with confidence in their accuracy.

Assignment-6: Simulating Forced Convection in a Confined Space

Objectives:
This assignment will use a computational fluid dynamics (CFD) program to model the mechanical-force-induced forced convection in a confined 2-D space.
Key learning point:

- Importance of boundary condition
- Influence of turbulence model.

Simulation Steps:

(1) Build a confined space with given dimensions as shown in Fig. 6.10;
(2) Prescribe proper boundary conditions including inlet, outlet, and four walls [iso-thermal case only: no temperature];
(3) Select a turbulence model: the standard k-ε model, and, the RNG k-ε model (or similar);
(4) Define convergence criterion: 0.1%;
(5) Set iteration: at least 2000 steps for steady simulation;
(6) Determine proper grid resolution with local refinement: at least 500,000 cells.

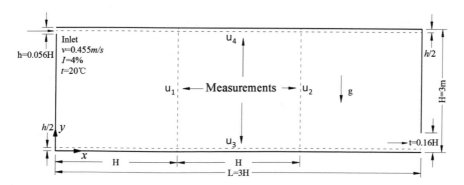

Fig. 6.10 Configuration, boundary condition setup and measurements of FC case

Cases to Be Simulated:

(1)   Test different outlet boundary condition settings (e.g., given V, P, or fully developed);
(2)   Test two turbulence models of choice using the most suitable outlet setting.

Report:

(1)   Case descriptions: description of the cases
(2)   Simulation details: computational domain, grid cells, convergence status

   • Figure of the grid used (on X-Y plane);
   • Figure of simulation convergence records.

(3)   Result and analysis

   • Figure of flow vectors;
   • Figure of pressure contours;
   • Figure of velocity contours;
   • Evaluate the influences of outlet boundary condition on simulation;
   • Evaluate the influences of turbulence model on simulation;
   • Validate the simulations with experimental data (Nielsen 1990).

(4)   Conclusions (findings, result implications, CFD experience and lessons, etc.)

References

Chen Q (1988) Indoor airflow, air quality and energy consumption of buildings. Ph.D. Thesis, Delft University of Technology, The Netherlands
Chen Q, Moser A (1991) Simulation of a multiple-nozzle diffuser. In Proceedings of the 12th AIVC conference on air movement and ventilation control within buildings, vol 2. Ottawa, Canada, pp 1–13
Launder BE, Sharma BI (1974) Application of the energy dissipation model of turbulence to the calculation of flow near a spinning disk. Lett Heat Mass Transf 1:131–138
Launder BE, Spalding DB (1974) The numerical computation of turbulent flows. Comput Methods Appl Mech Energy 3:269–289
Nielsen PV (1990) Specification of a two-dimensional test case. International energy agency
Rodi W (1991) Experience with two-layer models combining the K-epsilon model with a one-equation model near the wall. AIAA, Aerospace Sciences Meeting, 29th, Reno, NV, Jan. 7–10, 13 p
Srebric J, Chen Q (2002) Simplified numerical models for complex air supply diffusers. HVAC&R Res 8(3):277–294
Zhai Z, Chen Q (2004) Numerical determination and treatment of convective heat transfer coefficient in the coupled building energy and CFD simulation. Build Environ 39(8):1001–1009

Chapter 7
Generate Grid

7.1 Grid Classification

The fundamentals of CFD is to solve the continuous governing equations of fluid flows in a discrete numerical manner. Dividing the continuous flow domain into many discrete sub-domains (cells), hence, is a critical step in CFD, upon which the numerical governing equations can be resolved. The quality of discretized cells is crucial to the CFD simulation. Improper mesh may lead to divergence of simulation, inaccurate prediction, and/or slow convergence. Generating a grid (or mesh) of high quality is time-consuming even with the assistance of a commercial tool, often requesting a back-and-forth adjustment.

The continuous flow domain can be divided into small sub-domains (cells) in different methods. According to the yielded geometry of cells, the CFD grid can be classified into the following categories.

(1) *Rectangular versus Body-Fitted*

The rectangular grid is the simplest grid to generate, which divides the flow domain along the Cartesian coordinate directions X, Y, Z, respectively, into small ΔX_i, ΔY_j, ΔZ_k (i, j, k are the cell numbers in X, Y, Z direction). A 3-D cell (i, j, k) thus has a dimension of (ΔX_i, ΔY_j, ΔZ_k). Figure 7.1a shows a 2-D uniform rectangular grid with constant ΔX and ΔY. Using such a grid to simulate the flow around a solid (e.g., a square in Fig. 7.1a) requires assigning a few cells as solid (as shown in blue). If the geometry of the square aligns with the coordinate directions (Fig. 7.1a), the rectangular grid can catch the geometry precisely. However, if the square rotates at certain degrees (as shown in Fig. 7.1b), the exact geometry of the square can only be approximated because each cell must be identified as either solid or fluid depending on the size fraction of the two properties in the cell. The captured geometry accuracy can be improved if fine meshes are used around the object (Fig. 7.1c), while the artificial "resistance" (non-smoothness) at the square surface due to the discretization may still be observed. For some applications, this artificial "resistance" may be beneficial, representing the actual resistance effect that is not considered in

© Springer Nature Singapore Pte Ltd. 2020
Z. Zhai, *Computational Fluid Dynamics for Built
and Natural Environments*, https://doi.org/10.1007/978-981-32-9820-0_7

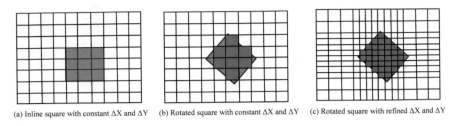

(a) Inline square with constant ΔX and ΔY   (b) Rotated square with constant ΔX and ΔY   (c) Rotated square with refined ΔX and ΔY

Fig. 7.1 Example of a 2-D rectangular grid for a flow around a square

the smooth surface model. Advanced numerical techniques were developed to fuzzily handle the cells around the irregular geometry to minimize the influences of the grid-caused non-smoothness. However, for most cases, a precise representation of true geometry is highly desired.

Body-fitted grid is commonly used to capture the exact geometries of irregular domains and objects. Both structured and unstructured grids can be generated to fit the geometries of flow domain, boundaries, and internal objects etc. The differences between structured and unstructured grids will be introduced next. Either approach demands sophisticated algorithms and significant efforts to generate a mesh with proper quality. Figure 7.2 illustrates 2-D body-fitted structured and unstructured grids for an airfoil. Table 7.1 summarizes the pros and cons of rectangular and body-fitted grids. As opposite to rectangular grid, body-fitted grid can precisely capture

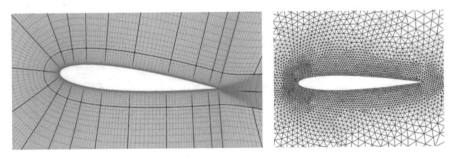

Fig. 7.2 2-D body-fitted structured (left) (Nordanger et al. 2015) and unstructured (right) (Barnosa Pola and Venturini Pola 2019) grids for an airfoil

Table 7.1 Comparison of pros and cons of rectangular and body-fitted grids		Rectangular grid	Body-fitted grid
	Capture complex geometries	Bad	Good
	Calculate geometry variables	Good	Bad
	Generate good-quality grids	Good	Bad

the irregular flow geometries, but it is somehow difficult to calculate the geometry related variables such as surface areas and volumes of irregular and curved cells. Most computer algorithms to automatically generate body-fitted grids cannot guarantee good quality of grid at critical places, such as near surfaces, which however are extremely important for CFD simulation. Manual adjustments are often requested that increase the simulation complexity and effort.

(2) *Structured versus Unstructured Grid*

The initial grid system developed for CFD simulation was structured grid, which has orderly data structure. Strictly speaking, a structured mesh can be recognized by all interior nodes of the mesh having an equal number of adjacent elements. Conventionally, the mesh generated by a structured grid generator is all quad or hexahedral. The structured grid is convenient for finite volume and finite difference analysis. It is usually easier to generate a structured grid than an unstructured grid for simple flow domains with regular objects. However, it is fairly challenging to produce a high-quality body-fitting structured grid, especially for complicated flow geometries. Unstructured grid was introduced to resolve this challenge. Unstructured grid is commonly used for finite element analysis (e.g., for solid mechanics), which has its own sophisticated data structure describing cells and vertexes. Unstructured grid using a variety of simple shapes such as triangles or tetrahedra is flexible to approach the irregular geometries of boundaries and objects.

Rectangular grid is a simplest form of structured grid, while body-fitted grid can be either structured or unstructured. Figure 7.2 left shows an example of body-fitted structured grid and right shows an example of body-fitted unstructured grid for a 2-D airfoil. Extra care and effort are required to produce the structured grid with perpendicular grids around the airfoil surface to ensure a convergent and accurate prediction (to be discussed later). This, however, can be readily addressed by the unstructured grid with triangles.

When developing a structured grid, the first step is to convert the physical domain (X, Y, Z) to the computational domain (i.e., numerical/logical domain) (i, j, k). All the physical variables, such as X, Y, Z, V, T, P, C etc., will be presented as a function of (i, j, k) (the discrete cell identify): $X(i, j, k)$, $Y(i, j, k)$, $Z(i, j, k)$, $V(i, j, k)$, $T(i, j, k)$, $P(i, j, k)$, $C(i, j, k)$. The algebraic operations will be performed on each cell (i, j, k). Figure 7.3 shows a 2-D physical flow domain (with boundary conditions) and its corresponding computational domain (and boundaries). Data array can be assigned in computer memory to store all the variables in the sweeping sequence of i, j, k (e.g., first through i, and then j and k for each variable). A one-dimensional array (called vector) is often used for CFD on vector machines that appeared in the early 1970s and dominated supercomputer design through the 1970s into the 1990s, as noted by various Cray platforms.

Rectangular structured grid is easy to generate by dividing X, Y, Z coordinates in the range of flow domain, independently, into small pieces ΔX_i, ΔY_j, ΔZ_k that can be either uniform or non-uniform (finer at flow regions with greater gradients of key variables). The formed rectangular cell of (i, j, k) with ΔX_i, ΔY_j, ΔZ_k has its surface areas of $\Delta X_i \times \Delta Y_j$, $\Delta Y_j \times \Delta Z_k$, $\Delta Z_k \times \Delta X_i$, respectively in Z, X,

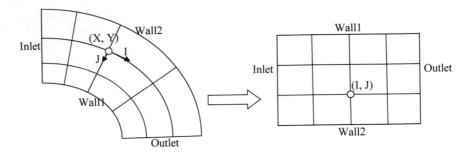

Fig. 7.3 Example of conversion from physical domain to computational domain

and Y directions, and its volume of $\Delta X_i \times \Delta Y_j \times \Delta Z_k$. A number of mathematical and numerical methods were developed to generate body-fitted structured grids, such as the complex variables techniques (Brown and Churchill 2014), various algebraic methods such as the transfinite interpolation method (Eiseman and Smith 1990), and elliptic partial differential equations methods (Thomas and Middlecoff 1980), etc.

To convert the governing equations in physical domain to those in structured computational domain, the following derivative transforms are operated:

$$\frac{\partial}{\partial x} = \frac{\partial}{\partial i}\frac{\partial i}{\partial x} + \frac{\partial}{\partial j}\frac{\partial j}{\partial x} + \frac{\partial}{\partial k}\frac{\partial k}{\partial x} \tag{7.1}$$

$$\frac{\partial}{\partial y} = \frac{\partial}{\partial i}\frac{\partial i}{\partial y} + \frac{\partial}{\partial j}\frac{\partial j}{\partial y} + \frac{\partial}{\partial k}\frac{\partial k}{\partial y} \tag{7.2}$$

$$\frac{\partial}{\partial z} = \frac{\partial}{\partial i}\frac{\partial i}{\partial z} + \frac{\partial}{\partial j}\frac{\partial j}{\partial z} + \frac{\partial}{\partial k}\frac{\partial k}{\partial z} \tag{7.3}$$

$$\begin{pmatrix} \frac{\partial}{\partial x} \\ \frac{\partial}{\partial y} \\ \frac{\partial}{\partial z} \end{pmatrix} = \begin{pmatrix} \frac{\partial i}{\partial x} & \frac{\partial j}{\partial x} & \frac{\partial k}{\partial x} \\ \frac{\partial i}{\partial y} & \frac{\partial j}{\partial y} & \frac{\partial k}{\partial y} \\ \frac{\partial i}{\partial z} & \frac{\partial j}{\partial z} & \frac{\partial k}{\partial z} \end{pmatrix} \begin{pmatrix} \frac{\partial}{\partial i} \\ \frac{\partial}{\partial j} \\ \frac{\partial}{\partial k} \end{pmatrix} \tag{7.4}$$

$$\begin{pmatrix} \frac{\partial i}{\partial x} & \frac{\partial j}{\partial x} & \frac{\partial k}{\partial x} \\ \frac{\partial i}{\partial y} & \frac{\partial j}{\partial y} & \frac{\partial k}{\partial y} \\ \frac{\partial i}{\partial z} & \frac{\partial j}{\partial z} & \frac{\partial k}{\partial z} \end{pmatrix} = \begin{pmatrix} \frac{\partial x}{\partial i} & \frac{\partial y}{\partial i} & \frac{\partial z}{\partial i} \\ \frac{\partial x}{\partial j} & \frac{\partial y}{\partial j} & \frac{\partial z}{\partial j} \\ \frac{\partial x}{\partial k} & \frac{\partial y}{\partial k} & \frac{\partial z}{\partial k} \end{pmatrix}^{-1} = adj(A)/|A| \tag{7.5}$$

where $J = |A| = \begin{vmatrix} \frac{\partial x}{\partial i} & \frac{\partial y}{\partial i} & \frac{\partial z}{\partial i} \\ \frac{\partial x}{\partial j} & \frac{\partial y}{\partial j} & \frac{\partial z}{\partial j} \\ \frac{\partial x}{\partial k} & \frac{\partial y}{\partial k} & \frac{\partial z}{\partial k} \end{vmatrix}$ is the volume of each cell (i, j, k). The surface areas

of each cell in x, y, z directions can also be projected to i, j, k, respectively, as shown below for the x direction as an example:

$$dA_{x,i} = |d\vec{y} \times d\vec{z}|_i = \left| \left(\frac{\partial y}{\partial i}di + \frac{\partial y}{\partial j}dj + \frac{\partial y}{\partial k}dk \right) \times \left(\frac{\partial z}{\partial i}di + \frac{\partial z}{\partial j}dj + \frac{\partial z}{\partial k}dk \right) \right|_i$$

$$= \left(\frac{\partial y}{\partial j}\frac{\partial z}{\partial k} - \frac{\partial y}{\partial k}\frac{\partial z}{\partial j} \right) dj dk \tag{7.6}$$

$$dA_{x,j} = |d\vec{y} \times d\vec{z}|_j = \left| \left(\frac{\partial y}{\partial i}di + \frac{\partial y}{\partial j}dj + \frac{\partial y}{\partial k}dk \right) \times \left(\frac{\partial z}{\partial i}di + \frac{\partial z}{\partial j}dj + \frac{\partial z}{\partial k}dk \right) \right|_j$$

$$= \left(\frac{\partial y}{\partial k}\frac{\partial z}{\partial i} - \frac{\partial y}{\partial i}\frac{\partial z}{\partial k} \right) di dk \tag{7.7}$$

$$dA_{x,k} = |d\vec{y} \times d\vec{z}|_k = \left| \left(\frac{\partial y}{\partial i}di + \frac{\partial y}{\partial j}dj + \frac{\partial y}{\partial k}dk \right) \times \left(\frac{\partial z}{\partial i}di + \frac{\partial z}{\partial j}dj + \frac{\partial z}{\partial k}dk \right) \right|_k$$

$$= \left(\frac{\partial y}{\partial i}\frac{\partial z}{\partial j} - \frac{\partial y}{\partial j}\frac{\partial z}{\partial i} \right) di dj \tag{7.8}$$

Similar equations can be obtained for calculating surface areas in y and z direction. Once a computational grid is generated, the surface areas and volumes of discrete cells can be calculated, which can then be used during the entire simulation.

Unstructured grid has the convenience of approximating irregular geometries with its variety of polyhedral cell formats. However, the data structure to store the cell info can get sophisticated. Figure 7.4a illustrates a simple 2-D triangle-based unstructured grid, which can be generated using some straightforward methods (e.g., connecting the middle points of the sides of each triangle) such as those shown in Fig. 7.4b. Table 7.2 presents two conventional data structures for the 2-D triangle grid shown in Fig. 7.4a. Obviously, this data structure will become very complicated when a 3-D unstructured grid is produced with a fine resolution. Table 7.3 provides a brief sum on the pros and cons of structured and unstructured grids.

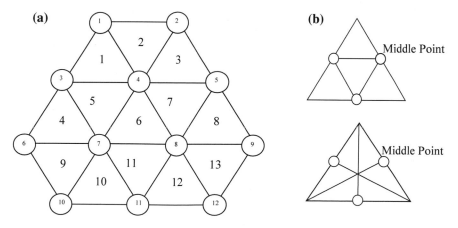

Fig. 7.4 A 2-D triangle unstructured grid and its data structure (**a**) and generation methods (**b**)

Table 7.2 Two conventional data structures for the 2-D unstructured grid shown in Fig. 7.4a

Method-1

Cell number	Vertex	Neighboring cell
1	$V_{11} = 1, V_{12} = 4,$ $V_{13} = 3$	$T_{11} = $ Null, $T_{12} = 2,$ $T_{13} = 5$
2	$V_{21} = 1, V_{22} = 2,$ $V_{23} = 4$	$T_{21} = 1, T_{22} = $ Null, $T_{23} = 3$
...	...	...
6	$V_{61} = 4, V_{62} = 8,$ $V_{63} = 7$	$T_{61} = 5, T_{62} = 7, T_{63}$ $= 11$
...	...	...

Method-2

Vertex	Connected vertex	Involved cell
1	$V_{11} = 2, V_{12} = 4, V_{13} = 3$	$E_{11} = 2, E_{12} = 1$
2	$V_{21} = 5, V_{22} = 4, V_{23} = 1$	$E_{21} = 3, E_{22} = 2$
...	...	...
7	$V_{71} = 3, V_{72} = 4, V_{73} = 8, V_{74} = 11,$ $V_{75} = 10, V_{76} = 6$	$E_{71} = 4, E_{72} = 5, E_{73} = 6, E_{74} = 11, E_{75}$ $= 10, E_{76} = 9$
...	...	...

Table 7.3 Comparison of pros and cons of structured and unstructured grids

	Structured grid	Unstructured grid
Capture complex geometries	Bad	Good
Data structures	Good	Bad

(3) *Staggered versus Non-staggered*

Staggered and non-staggered grids may show the exact same grid format. However, they store the discrete variables at different locations. Staggered grid was originally proposed in CFD with its explicit physical meaning and mathematical convenience. In a staggered grid, the pressure (and temperature, concentration etc.) information is stored at the cell center (e.g., P, W, E, S, N) while the velocity information is saved to the faces (e.g. w, e, s, n) of the cell (Fig. 7.5). Under this arrangement, the two adjacent pressure nodes directly appear in the discretized momentum equation, becoming the driving force of the flow. The in and out flow rates of various variables can also be directly computed with face velocities. The simulation based on a staggered grid system is robust and effective in removing the inherent problems associated with the pressure term and the continuity equation. However, this kind of grid system needs two sets of data structure to store the results and introduces extra complexity to the computation. For instance, face areas need be calculated not only at faces (e.g., w, e, s, n) but also at centers (e.g. P, W, E, S, N).

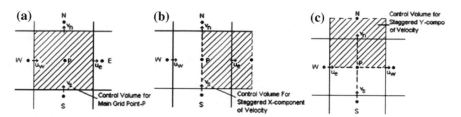

Fig. 7.5   Illustration of staggered grid and control volumes for pressure and velocities

Non-staggered (collocated) grid was introduced to reduce programing and computing sophistication, which becomes popular for most today's CFD programs (especially with unstructured grid). In a non-staggered grid, all the variables are stored at the discrete computation nodes (e.g., the central points of cells, i.e., P, W, E, S, N, B, T). In order to obtain the flow driving force from pressure difference, a linear interpolation method is often applied to calculate the pressures at cell faces (w, e, s, n, b, t) as demonstrated in Fig. 7.6. This practice usually leads to the non-physical oscillation or the so-called red-black checkerboard splitting of the pressure field and associated difficulties in obtaining a converged solution. As shown in Fig. 7.6 as an example, the predicted velocity at W, P, and E, respectively, can meet both mass and momentum conservations. From the momentum equation, U_W is positive driven by $P_{WW} - P_P = 2$ Pa and U_E is also positive with $P_P - P_{EE} = 2$ Pa, but U_P is negative as driven by $P_W - P_E = -2$ Pa. The face velocities at w (between W and P) and e (between P and E) are both zero, satisfying the mass conservation. However, the obtained result is not physical with the sudden velocity switch at adjacent points (and flow velocity at node while zero at faces).

A few solutions were proposed, and one widely used solution to avoid checkerboard splitting for cell-centered arrangement was to use the so-called momentum interpolation method (MIM) (Rhie and Chow 1983) to evaluate cell face variables

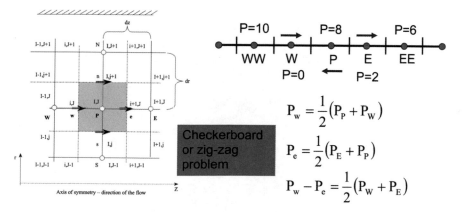

Fig. 7.6   Checkerboard (zig-zag) problem in non-staggered grid

Table 7.4 Comparison of
pros and cons of staggered
and non-staggered grids

	Staggered grid	Non-staggered grid
Simulation stability	Good	Bad
Programming effort	Bad	Good

from the cell centered quantities. This method will be introduced in detail in Chap. 8 when describing the velocity-pressure decoupling algorithms. Table 7.4 provides a brief summary on the pros and cons of staggered and non-staggered grids.

7.2   Advanced Grid Systems

(1)   *Local-refined grid/Multi-grid*

When generating a proper grid, fine resolution is inevitable to capture significant gradients of key variables (e.g., P, V, T, C), such as near rigid surfaces. A coarse grid will lead to inaccuracy of prediction and provide less useful information for process. Figure 7.7 shows an example of a refined body-fitted structured grid for a 3-D airfoil. As revealed, in order to capture the flow details at the trailing edge of the airfoil, the grid was refined in the solid-line circulated area; however, due to the structured nature of the grid, unnecessary refinement was yielded in the dash-line circulated areas, which increases the computing cost.

Local-refined grid (or multi-grid) provides a remedy to this problem, which only refines the grid for the needed areas. Figure 7.8 presents two examples of local-refined (or multi) grids. The left case shows a refinement by dividing the existing coarse grids into smaller cells, while the right case has multiple independent grids overlapping on each other. Obviously, independent grids are easy to generate with no constraint from the other grids. The technique challenge is to correctly and stably exchange the flow field information at the interfaces of different grids (or refinements)

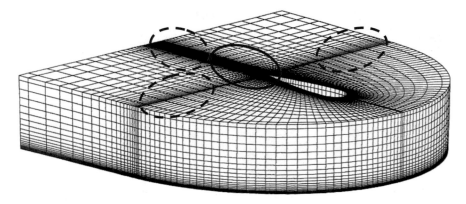

Fig. 7.7 Refined body-fitted structured grid for a 3-D airfoil

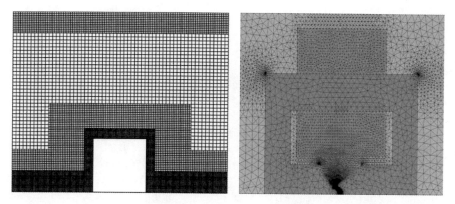

Fig. 7.8 Examples of local-refined grids (left: Cevheri et al. 2016; right: Römer et al. 2017)

during the numerical iteration. Typically, information transferring from a fine grid to a coarse grid can be realized by averaging the inputs of neighboring fine grid nodes. Interpolation techniques should be applied to distribute the coarse grid information to the fine grids, which can affect the prediction convergence, accuracy and efficiency. Applying multi-grids with the alternation between coarse and fine grids has been an effective approach to accelerating the CFD simulation.

(2) *Multi-block grid*

For complicated flow domains, generating one proper grid for the entire domain is very challenging, especially for structured body-fitted grids. One alternative approach is to divide the domain into blocks, with separate grids generated for each block, and then link the blocks with proper communications at the block interfaces. Figure 7.9 presents a branch flow that is almost impossible to obtain one structured grid. By

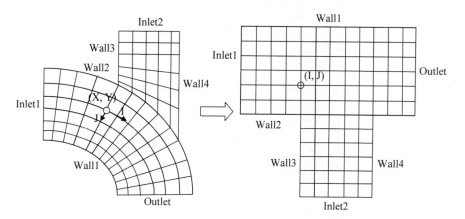

Fig. 7.9 Multi-block grid with a continuous patched interface for a branch flow in both physical and computational domain

dividing the flow domain into two blocks, separate grids of high quality can be easily produced. Certainly, two sets of (i, j) index must be used and stored, one for each block. Data will be transferred between the responding interface (i, j) cells at each grid (i.e., artificial internal boundary conditions for each block grid). Appropriate data communication/exchange algorithm determines the simulation correctness, accuracy, stability, and convergence.

The data exchange methods at the interfaces of blocks can be categorized as: the patched method and the overlapped method, each with continuous and discontinuous grid algorithms. The case shown in Fig. 7.9 has a patched continuous grid at the interface. If the two blocks have discontinuous grids at the interface, it becomes a patched discontinuous grid case. For the patched cases, flux conservation of various variables should be ensured between the neighboring blocks. The patched method has a good accuracy, stability and convergence, but is somehow complicated to implement in the code. Figure 7.10 shows an example with an overlapped interface of two blocks. The overlapped interface has the discontinuous grids. For cases with an overlapped continuous interface grid, the data from two blocks can be directly exchanged at the same nodes. For cases with discontinuous grids from own blocks at the interface such as that in Fig. 7.10, interpolation is required to transfer information between the nodes from two grids. In general, a lower-order interpolation algorithm can provide a good accuracy and stability. The overlapped method is relatively easy to program in CFD.

(3) *Adaptive grid*

Most CFD simulations utilize a same grid during the entire modeling iteration. Grid refinement is pre-assigned for critical locations such as near inlets, outlets, and surfaces as well as the regions that are expected to have large gradients of key variables (e.g., vortices behind a solid object etc.). However, for some cases, the flow structures are unpredictable before the simulation. Adaptive grid becomes a valuable approach to handling the scenarios. Starting from a uniform and coarse grid, the "static" adaptive grid method will increase the total grid number by dividing large cells at critical locations into smaller cells at each simulation step after evaluating the gradients of

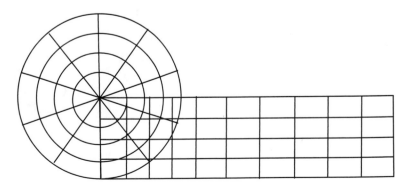

Fig. 7.10 Multi-block grid with a discontinuous overlapped interface

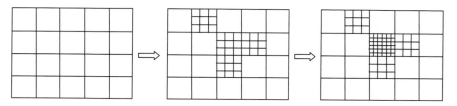

Fig. 7.11 A typical evolution process of a static adaptive grid

the key variables of interest. Figure 7.11 demonstrates a typical evolution process of a static adaptive grid. The "static" adaptive grid method can begin with a fast simulation and gradually refine the grid wherever is necessary. It is thus very efficient. However, the addition of new cells/nodes at "random" locations imposes challenges to the data structure management and further the simulation process. Figure 7.12 presents the final adaptive grid for a flow around turbine blades. The auto-refined (adaptive) grid explicitly reveals the locations of the shock waves that are critical for the analysis.

The "dynamic" adaptive grid method keeps the total grid number as a constant (so as the data structure). The cell/node positions will be shifted during the iteration by evaluating the gradients of the key variables at each step. Cells/nodes will be moved towards the areas with larger gradients of the variables. Figure 7.13 shows one example of resultant adaptive grids with different shifting weighting factors.

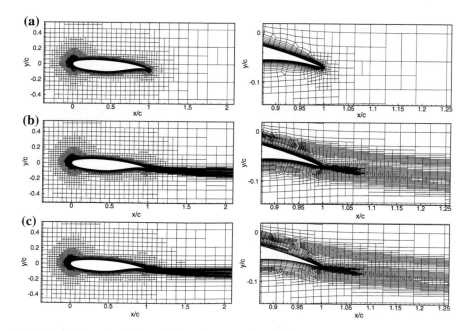

Fig. 7.12 Final static adaptive grid for a flow around an airfoil (Wackers et al. 2017)

Fig. 7.13 Example of resultant dynamic adaptive grids with different shifting factors

Obviously, the benefit of easy data structure is compensated by the distorted cell shape (i.e., cell quality). A moderately fine grid is often required from the beginning of a simulation as the total grid number is unchanged, which increases the computing cost as compared with the "static" adaptive grid method.

7.3   Grid Quality Control

CFD simulation is performed on discrete grid (cells). Quality of numerical grid has a direct impact on CFD prediction in terms of correctness, accuracy, convergence, stability, and speed. The following three criteria are important to judge the quality of a grid:

- Fine grid at large gradient
- Size ratio of adjacent cells (smoothness)
- Skewness.

   As stated before, fine resolution is expected at locations where large gradients of key variables exist in order to capture the flow details and avoid numerical errors (due to interpolation). Typically, these regions require refined grids: rigid surfaces (e.g., wall, internal object), charge/discharge openings (e.g., window, jet), and various sources of mass, momentum and energy.

   The cell size change should be gradual and sudden size jump between adjacent cells should be avoided as illustrated in Fig. 7.14. Continuous size change is visually represented by the smoothness of the numerical CFD model to be close to the actual physical model. Large cell size difference also leads to the interpolation of face values leaning on the influence of only one side but less contribution from the other side. This will result in a slower convergence and less accuracy of simulation. In general, less than 20% difference of size ratio between adjacent cells is suggested, i.e., $0.8 \leq \Delta X_{i+1}/\Delta X_i \leq 1.2$ for each direction.

   A skew grid results in large error in numerical approximation, which is especially important for critical areas with complex turbulence physics such as near the walls. Most numerical treatments of boundary conditions assume orthogonal or equiangular

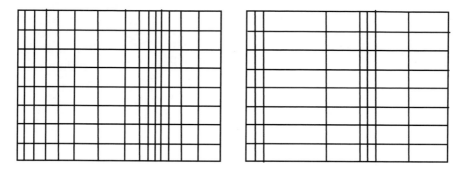

Fig. 7.14 Grid cell size change (left) gradual (right) sudden

conditions, such as wall functions. Significant skewness from the ideal conditions leads to considerable simulation inaccuracy, instability, and divergence, etc.

Avoiding grid skewness for structured grids imply:

(1) keeping orthogonality of cell (rectangular prism);
(2) keeping reasonable aspect ratio of cell (ideally cube if flow is multi-directional);
(3) keeping one coordinate perpendicular to the boundaries (inlet, outlet, and wall etc.).

Generating a proper structured grid that can meet all these conditions is challenging, especially for irregular flow domains and object geometries. Trade-off is inevitable, which heavily counts on the experience of a user, with a good understanding of the involved flow physics, the used numerical methods, the CFD program, and the project expectations.

Two methods are commonly utilized to quantitatively measure skewness of unstructured grids:

• Based on the equilateral volume (applies only to triangles and tetrahedra);
• Based on the deviation from a normalized equilateral angle (applies to all cell and face shapes, e.g., pyramids and prisms).

In the equilateral volume deviation method, the skewness is defined as:

$$Skewness = \frac{Optimal\ cell\ size - actual\ cell\ size}{Optimal\ cell\ size} \tag{7.9}$$

where the optimal cell size is the size of an equilateral cell with the same circumradius (Fig. 7.15).

In the normalized angle deviation method, skewness is defined as:

$$Skewness = \max\left[\frac{\theta_{max} - \theta_e}{180 - \theta_e}, \frac{\theta_e - \theta_{min}}{\theta_e}\right] \tag{7.10}$$

Fig. 7.15 Illustration of
optimal cell and actual cell
with the same circumradius

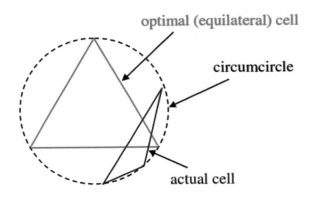

Table 7.5 Range of skewness values and the corresponding cell quality

Value of Skewness	0–0.25	0.25–0.50	0.50–0.80	0.80–0.95	0.95–0.99	0.99–1.00
Cell quality	Excellent	Good	Acceptable	Poor	Sliver	Degenerate

where: θ_{max} = largest angle in the face or cell; θ_{min} = smallest angle in the face or cell; θ_e = angle for an equiangular face/cell (e.g., 60 for a triangle, 90 for a square). For a pyramid, the cell skewness will be the maximum skewness computed for any face. An ideal pyramid (skewness = 0) is one in which the four triangular faces are equilateral (and equiangular) and the quadrilateral base face is a square.

Table 7.5 lists the range of skewness values and the corresponding cell quality. According to the definition of skewness, a value of 0 indicates an equilateral cell (best) and a value of 1 indicates a completely degenerate cell (worst). Degenerate cells (slivers) are characterized by nodes that are nearly coplanar (colinear in 2D). Highly skewed faces and cells are unacceptable because the equations being solved assume that the cells are relatively equilateral/equiangular. If bad cells are found during the grid generation, these bad cells should be deleted, and necessary decomposition should be performed with pre-meshing edges and faces for re-meshing.

7.4   Numerical Viscosity

Discretizing a continuous spatial and/or temporal domain introduces numerical errors to the exact solution of the flow governing equations. Finer discretization resolution leads to less numerical error. Theoretically, when refining grid resolutions, there is a point that further refinement will not (or only slightly) change numerical solutions. This is the minimum grid resolution upon which a grid-independent solution can be obtained. The truncation error caused by discretization of the governing equations of fluid flow is the fundamental reason for this. The following sections quantitatively assess the truncation error brought into numerical solutions of the governing equations in the form of numerical (artificial) viscosity. The analyses are based on the momentum equations of instantaneous velocities, while the same conclusions can be applied to RANS based governing equations, as well as temperature and concentration equations.

(1) Upwind Differencing Scheme

First order upwind scheme is simple, popular and unconditionally stable; however, it may lead to physically invalid results in some applications. Using a two-dimensional (2D) steady-state incompressible flow as an example, the momentum equations are expressed as:

$$
\begin{cases}
u\frac{\partial u}{\partial x} + v\frac{\partial u}{\partial y} = -\frac{1}{\rho}\frac{\partial P}{\partial x} + v\left[\frac{\partial^2 u}{\partial x^2} + \frac{\partial^2 u}{\partial y^2}\right] + S_x \\
u\frac{\partial v}{\partial x} + v\frac{\partial v}{\partial y} = -\frac{1}{\rho}\frac{\partial P}{\partial y} + v\left[\frac{\partial^2 v}{\partial x^2} + \frac{\partial^2 v}{\partial y^2}\right] + S_y
\end{cases}
\tag{7.11}
$$

where u and v are the velocity components along x and y direction, P is the pressure of the fluid, ρ and v are the density and physical kinematic viscosity, and S_x and S_y are the external forces on the fluid along x and y direction. Using the upwind numerical scheme, assume $u > 0$ and $v > 0$, the 1st order term can be discretized as:

$$
\begin{cases}
\frac{\partial u}{\partial x} = \frac{u_{i,j}-u_{i-1,j}}{\Delta x} \\
\frac{\partial u}{\partial y} = \frac{u_{i,j}-u_{i,j-1}}{\Delta y} \\
\frac{\partial v}{\partial x} = \frac{v_{i,j}-v_{i-1,j}}{\Delta x} \\
\frac{\partial v}{\partial y} = \frac{v_{i,j}-v_{i,j-1}}{\Delta y}
\end{cases}
\tag{7.12}
$$

The steady-state momentum equations can thus be discretized as:

$$
\begin{cases}
u\frac{u_{i,j}-u_{i-1,j}}{\Delta x} + v\frac{u_{i,j}-u_{i,j-1}}{\Delta y} = -\frac{1}{\rho}\frac{\partial P}{\partial x} + v\left[\frac{\partial^2 u}{\partial x^2} + \frac{\partial^2 u}{\partial y^2}\right] + S_x \\
u\frac{v_{i,j}-v_{i-1,j}}{\Delta x} + v\frac{v_{i,j}-v_{i,j-1}}{\Delta y} = -\frac{1}{\rho}\frac{\partial P}{\partial y} + v\left[\frac{\partial^2 v}{\partial x^2} + \frac{\partial^2 v}{\partial y^2}\right] + S_y
\end{cases}
\tag{7.13}
$$

Using the Taylor series to express the variables on the adjacent cells:

$$
\begin{cases}
u_{i-1,j} = u_{i,j} - \Delta x\frac{\partial u_{i,j}}{\partial x} + \frac{\Delta x^2}{2}\frac{\partial^2 u_{i,j}}{\partial x^2} + O\left(\Delta x^3\right) \\
u_{i,j-1} = u_{i,j} - \Delta y\frac{\partial u_{i,j}}{\partial y} + \frac{\Delta y^2}{2}\frac{\partial^2 u_{i,j}}{\partial y^2} + O\left(\Delta y^3\right)
\end{cases}
\tag{7.14}
$$

This yields:

$$
\begin{cases}
\frac{u_{i,j}-u_{i-1,j}}{\Delta x} = \frac{\partial u_{i,j}}{\partial x} - \frac{\Delta x}{2}\frac{\partial^2 u_{i,j}}{\partial x^2} - O\left(\Delta x^2\right) \\
\frac{u_{i,j}-u_{i,j-1}}{\Delta y} = \frac{\partial u_{i,j}}{\partial y} - \frac{\Delta y}{2}\frac{\partial^2 u_{i,j}}{\partial y^2} - O\left(\Delta y^2\right)
\end{cases}
\tag{7.15}
$$

Similarly,

$$\begin{cases} v_{i-1,j} = v_{i,j} - \Delta x \frac{\partial v_{i,j}}{\partial x} + \frac{\Delta x^2}{2}\frac{\partial^2 v_{i,j}}{\partial x^2} + O(\Delta x^3) \\ v_{i,j-1} = v_{i,j} - \Delta y \frac{\partial v_{i,j}}{\partial y} + \frac{\Delta y^2}{2}\frac{\partial^2 v_{i,j}}{\partial y^2} + O(\Delta y^3) \end{cases} \tag{7.16}$$

and

$$\begin{cases} \frac{v_{i,j}-v_{i-1,j}}{\Delta x} = \frac{\partial v_{i,j}}{\partial x} - \frac{\Delta x}{2}\frac{\partial^2 v_{i,j}}{\partial x^2} - O(\Delta x^2) \\ \frac{v_{i,j}-v_{i,j-1}}{\Delta y} = \frac{\partial v_{i,j}}{\partial y} - \frac{\Delta y}{2}\frac{\partial^2 v_{i,j}}{\partial y^2} - O(\Delta y^2) \end{cases} \tag{7.17}$$

Substituting the discretization terms in Eq. (7.13) with the Taylor series expansions (7.15) and (7.17), the momentum equations become:

$$\begin{cases} u\left(\frac{\partial u_{i,j}}{\partial x} - \frac{\Delta x}{2}\frac{\partial^2 u_{i,j}}{\partial x^2} - O(\Delta x^2)\right) + v\left(\frac{\partial u_{i,j}}{\partial y} - \frac{\Delta y}{2}\frac{\partial^2 u_{i,j}}{\partial y^2} - O(\Delta y^2)\right) = -\frac{1}{\rho}\frac{\partial P}{\partial x} + v\left[\frac{\partial^2 u}{\partial x^2} + \frac{\partial^2 u}{\partial y^2}\right] + S_x \\ u\left(\frac{\partial v_{i,j}}{\partial x} - \frac{\Delta x}{2}\frac{\partial^2 v_{i,j}}{\partial x^2} - O(\Delta x^2)\right) + v\left(\frac{\partial v_{i,j}}{\partial y} - \frac{\Delta y}{2}\frac{\partial^2 v_{i,j}}{\partial y^2} - O(\Delta y^2)\right) = -\frac{1}{\rho}\frac{\partial P}{\partial y} + v\left[\frac{\partial^2 v}{\partial x^2} + \frac{\partial^2 v}{\partial y^2}\right] + S_y \end{cases} \tag{7.18}$$

and further as:

$$\begin{cases} u\frac{\partial u}{\partial x} + v\frac{\partial u}{\partial y} = -\frac{1}{\rho}\frac{\partial P}{\partial x} + \frac{u\cdot\Delta x}{2}\frac{\partial^2 u}{\partial x^2} + \frac{v\cdot\Delta y}{2}\frac{\partial^2 u}{\partial y^2} + v\left[\frac{\partial^2 u}{\partial x^2} + \frac{\partial^2 u}{\partial y^2}\right] + S_x + O(\Delta x^2) + O(\Delta y^2) \\ u\frac{\partial v}{\partial x} + v\frac{\partial v}{\partial y} = -\frac{1}{\rho}\frac{\partial P}{\partial y} + \frac{u\cdot\Delta x}{2}\frac{\partial^2 v}{\partial x^2} + \frac{v\cdot\Delta y}{2}\frac{\partial^2 v}{\partial y^2} + v\left[\frac{\partial^2 v}{\partial x^2} + \frac{\partial^2 v}{\partial y^2}\right] + S_y + O(\Delta x^2) + O(\Delta y^2) \end{cases} \tag{7.19}$$

Since the coefficients $(u \cdot \Delta x)/2$ and $(v \cdot \Delta y)/2$ have the same effect as the physical viscosity of the fluid, it is called artificial viscosity [90] or numerical viscosity. The numerical viscosities for x and y directions with the first order upwind scheme, respectively, are:

$$\begin{cases} v_x = u\frac{\Delta x}{2} \\ v_y = v\frac{\Delta y}{2} \end{cases} \tag{7.20}$$

If $u < 0$, $v_x = -u\,\Delta x/2$, and thus $v_x = |u\,\Delta x/2|$. If $v < 0$, $v_y = -v\,\Delta y/2$, and thus $v_y = |v\,\Delta y/2|$. Similarly, for 3D cases, $v_z = |w\,\Delta z/2|$. The magnitude of numerical viscosity is proportional to the grid size, which reveals the fact that refining grid improves the accuracy of a CFD simulation.

For a hybrid differencing scheme, the upwind discretization is employed when the Péclet number is greater than 2 and the central differential scheme (CDS) is used for Pe $\leq$ 2 [91]. The detailed derivation of associated numerical viscosity term is as follows.

(2) *Central Differencing Scheme (CDS)*

CDS is the scheme often used by the Hybrid scheme when $Pe \leq 2$. The expression of the Hybrid scheme is:

$$u_i = \begin{cases} u_{i-1} & Pe > 2\,upwind\,scheme \\ \frac{1}{2}(u_{i-1} + u_{i+1}) & Pe \leq 2\,central\,scheme \end{cases} \tag{7.21}$$

For CDS, assume u > 0 and v > 0, the 1st order derivative term can be discretized as:

$$\frac{\partial u}{\partial x} = \frac{u_{i+1,j} - u_{i-1,j}}{2\Delta x} \tag{7.22}$$

$$\frac{\partial u}{\partial y} = \frac{u_{i,j+1} - u_{i,j-1}}{2\Delta y} \tag{7.23}$$

$$\frac{\partial v}{\partial x} = \frac{v_{i+1,j} - v_{i-1,j}}{2\Delta x} \tag{7.24}$$

$$\frac{\partial v}{\partial y} = \frac{v_{i,j+1} - v_{i,j-1}}{2\Delta y} \tag{7.25}$$

The steady-state momentum equation can be discretized accordingly as:

$$u\frac{u_{i+1,j} - u_{i-1,j}}{2\Delta x} + v\frac{u_{i,j+1} - u_{i,j-1}}{2\Delta y} = -\frac{1}{\rho}\frac{\partial P}{\partial x} + v\left[\frac{\partial^2 u}{\partial x^2} + \frac{\partial^2 u}{\partial y^2}\right] + S_x \tag{7.26}$$

$$u\frac{v_{i+1,j} - v_{i-1,j}}{2\Delta x} + v\frac{v_{i,j+1} - v_{i,j-1}}{2\Delta y} = -\frac{1}{\rho}\frac{\partial P}{\partial y} + v\left[\frac{\partial^2 v}{\partial x^2} + \frac{\partial^2 v}{\partial y^2}\right] + S_y \tag{7.27}$$

Using the Taylor series to express the variables on the adjacent cells:

$$u_{i+1,j} = u_{i,j} + \Delta x\frac{\partial u_{i,j}}{\partial x} + \frac{\Delta x^2}{2}\frac{\partial^2 u_{i,j}}{\partial x^2} + \frac{\Delta x^3}{6}\frac{\partial^3 u_{i,j}}{\partial x^3} + O(\Delta x^4) \tag{7.28}$$

$$u_{i-1,j} = u_{i,j} - \Delta x\frac{\partial u_{i,j}}{\partial x} + \frac{\Delta x^2}{2}\frac{\partial^2 u_{i,j}}{\partial x^2} - \frac{\Delta x^3}{6}\frac{\partial^3 u_{i,j}}{\partial x^3} + O(\Delta x^4) \tag{7.29}$$

and

$$u_{i,j+1} = u_{i,j} + \Delta y\frac{\partial u_{i,j}}{\partial y} + \frac{\Delta y^2}{2}\frac{\partial^2 u_{i,j}}{\partial y^2} + \frac{\Delta y^3}{6}\frac{\partial^3 u_{i,j}}{\partial y^3} + O(\Delta y^4) \tag{7.30}$$

$$u_{i,j-1} = u_{i,j} - \Delta y\frac{\partial u_{i,j}}{\partial y} + \frac{\Delta y^2}{2}\frac{\partial^2 u_{i,j}}{\partial y^2} - \frac{\Delta y^3}{6}\frac{\partial^3 u_{i,j}}{\partial y^3} + O(\Delta y^4) \tag{7.31}$$

Therefore

$$\frac{u_{i+1,j} - u_{i-1,j}}{2\Delta x} = \frac{\partial u_{i,j}}{\partial x} + \frac{\Delta x^2}{6}\frac{\partial^3 u_{i,j}}{\partial x^3} + O(\Delta x^3) \tag{7.32}$$

$$\frac{u_{i,j+1} - u_{i,j-1}}{2\Delta y} = \frac{\partial u_{i,j}}{\partial y} + \frac{\Delta y^2}{6}\frac{\partial^3 u_{i,j}}{\partial y^3} + O(\Delta y^3) \tag{7.33}$$

Similarly

$$\frac{v_{i+1,j} - v_{i-1,j}}{2\Delta x} = \frac{\partial v_{i,j}}{\partial x} + \frac{\Delta x^2}{6} \frac{\partial^3 v_{i,j}}{\partial x^3} + O(\Delta x^3) \tag{7.34}$$

$$\frac{v_{i,j+1} - v_{i,j-1}}{2\Delta y} = \frac{\partial v_{i,j}}{\partial y} + \frac{\Delta y^2}{6} \frac{\partial^3 v_{i,j}}{\partial y^3} + O(\Delta y^3) \tag{7.35}$$

Substituting the discretization terms in Eqs. (7.26) and (7.27) with the Taylor series expansions (7.32)–(7.35) provides:

$$u\left(\frac{\partial u_{i,j}}{\partial x} + \frac{\Delta x^2}{6} \frac{\partial^3 u_{i,j}}{\partial x^3} + O(\Delta x^3)\right) + v\left(\frac{\partial u_{i,j}}{\partial y} + \frac{\Delta y^2}{6} \frac{\partial^3 u_{i,j}}{\partial y^3} + O(\Delta y^3)\right)$$
$$= -\frac{1}{\rho}\frac{\partial P}{\partial x} + v\left[\frac{\partial^2 u}{\partial x^2} + \frac{\partial^2 u}{\partial y^2}\right] + S_x \tag{7.36}$$

$$u\left(\frac{\partial v_{i,j}}{\partial x} + \frac{\Delta x^2}{6} \frac{\partial^3 v_{i,j}}{\partial x^3} + O(\Delta x^3)\right) + v\left(\frac{\partial v_{i,j}}{\partial y} + \frac{\Delta y^2}{6} \frac{\partial^3 v_{i,j}}{\partial y^3} + O(\Delta y^3)\right)$$
$$= -\frac{1}{\rho}\frac{\partial P}{\partial y} + v\left[\frac{\partial^2 v}{\partial x^2} + \frac{\partial^2 v}{\partial y^2}\right] + S_y \tag{7.37}$$

Further as:

$$u\frac{\partial u}{\partial x} + v\frac{\partial u}{\partial y} = -\frac{1}{\rho}\frac{\partial P}{\partial x} - \frac{u \cdot \Delta x^2}{6}\frac{\partial^3 u}{\partial x^3} - \frac{v \cdot \Delta y^2}{6}\frac{\partial^3 u}{\partial y^3} + v\left[\frac{\partial^2 u}{\partial x^2} + \frac{\partial^2 u}{\partial y^2}\right]$$
$$+ S_x + O(\Delta x^3) + O(\Delta y^3) \tag{7.38}$$

$$u\frac{\partial v}{\partial x} + v\frac{\partial v}{\partial y} = -\frac{1}{\rho}\frac{\partial P}{\partial y} - \frac{u \cdot \Delta x^2}{6}\frac{\partial^3 v}{\partial x^3} - \frac{v \cdot \Delta y^2}{6}\frac{\partial^3 v}{\partial y^3} + v\left[\frac{\partial^2 v}{\partial x^2} + \frac{\partial^2 v}{\partial y^2}\right]$$
$$+ S_y + O(\Delta x^3) + O(\Delta y^3) \tag{7.39}$$

In order to have the same format as physical viscosity, the numerical viscosities for x and y directions are expressed as coefficients of the second derivative term, respectively, as:

$$v_x = \frac{\frac{-u \cdot \Delta x^2}{6}\frac{\partial^3 u}{\partial x^3}}{\frac{\partial^2 u}{\partial x^2}} \tag{7.40}$$

$$v_y = \frac{\frac{-v \cdot \Delta y^2}{6}\frac{\partial^3 u}{\partial y^3}}{\frac{\partial^2 u}{\partial y^2}} \tag{7.41}$$

If u < 0

$$v_x = \frac{\frac{u \cdot \Delta x^2}{6}\frac{\partial^3 u}{\partial x^3}}{\frac{\partial^2 u}{\partial x^2}} \tag{7.42}$$

If $v < 0$

$$v_y = \frac{\dfrac{v \cdot \Delta y^2}{6} \dfrac{\partial^3 u}{\partial y^3}}{\dfrac{\partial^2 u}{\partial y^2}} \tag{7.43}$$

Therefore, a general expression of the numerical viscosity term is (on U equation):

$$v_x = -\frac{\dfrac{|u| \cdot \Delta x^2}{6} \dfrac{\partial^3 u}{\partial x^3}}{\dfrac{\partial^2 u}{\partial x^2}} \tag{7.44}$$

$$v_y = -\frac{\dfrac{|v| \cdot \Delta y^2}{6} \dfrac{\partial^3 u}{\partial y^3}}{\dfrac{\partial^2 u}{\partial y^2}} \tag{7.45}$$

Applying the same approach to the z direction yields

$$v_z = -\frac{\dfrac{|w| \cdot \Delta z^2}{6} \dfrac{\partial^3 u}{\partial z^3}}{\dfrac{\partial^2 u}{\partial z^2}} \tag{7.46}$$

Hence, a general expression of numerical viscosity of CDS on u_i equation is (without the summation rule):

$$v_{i,j} = -\frac{\dfrac{|u_j| \cdot \Delta x_j^2}{6} \dfrac{\partial^3 u_i}{\partial x_j^3}}{\dfrac{\partial^2 u_i}{\partial x_j^2}} \tag{7.47}$$

Similar analysis can be performed for other differencing schemes such as QUICK. Table 7.6 summarises the numerical viscosities and truncation errors from discretization with different numerical schemes. A grid independent solution theoretically

Table 7.6 Numerical viscosities and truncation errors from discretization with different numerical schemes

Scheme	Numerical viscosity	Truncation error	Higher order term (H.O.T)	Note
Upwind	$\left| u_j \dfrac{\Delta x_j}{2} \right|$	$\left| u_j \dfrac{\Delta x_j}{2} \right| \dfrac{\partial^2 u_i}{\partial x_j^2}$	$O(\Delta x^2)$	Same as hybrid when $Pe > 2$
CDS	$-\dfrac{\dfrac{|u_j| \cdot \Delta x_j^2}{6} \dfrac{\partial^3 u_i}{\partial x_j^3}}{\dfrac{\partial^2 u_i}{\partial x_j^2}}$	$-\dfrac{|u_j| \cdot \Delta x_j^2}{6} \dfrac{\partial^3 u_i}{\partial x_j^3}$	$O(\Delta x^3)$	Same as hybrid when $Pe \le 2$
Quick	$-\dfrac{\dfrac{|u_j| \cdot \Delta x_j^2}{24} \dfrac{\partial^3 u_i}{\partial x_j^3}}{\dfrac{\partial^2 u_i}{\partial x_j^2}}$	$-\dfrac{|u_j| \cdot \Delta x_j^2}{24} \dfrac{\partial^3 u_i}{\partial x_j^3}$	$O(\Delta x^3)$	

requires the numerical viscosity to be much smaller than the actual viscosity (either laminar or turbulent viscosity), so that the effect of grid-induced-error is fully eliminated from the numerically solved governing equations. Since numerical viscosity is determined by both grid size and local velocities (derivatives), it is difficult to find a uniformly suitable grid size that can always meet the requirement of grid independency for various simulation conditions. Nevertheless, the numerical viscosity provides an important aspect to evaluate analytically whether grid independency is reached or not.

7.5 Applications of Structured Coarse Grids

Refinement of grid can reduce the impact of discretization-induced numerical error and thus improve the accuracy of CFD simulation. Adopting high-order discretizing schemes can be another potential solution, which however may impose convergence and stability challenges.

Reaching a grid-independent simulation solution is the first task in CFD before the predicted results can be analyzed. Chapter 9 will introduce the procedure and indices for attaining the grid-independent numerical solution, while this chapter focuses on generating various appropriate grids. Grid-independent solutions demand a significantly large number of grid cells, challenging both computational resources and skills. It also prohibits the application of CFD for most engineering problems. Obtaining not perfect but reasonable CFD results with moderate (or even coarse) grids are highly desired for assisting engineering design, evaluation and control tasks.

Wang and Zhai (2012) investigated the grid-induced errors, evaluated the potential computing cost saving by using coarse grids, and provided a guideline for optimizing the trade-off between grid resolution and computing cost. The study indicated that the numerical error caused by coarse grid can be minimized by properly adapting the distribution of grid size. Following the guideline of coarse-grid specifications, coarse-grid CFD can provide informative prediction that is comparable to a grid-independent result. The computing cost of CFD with an optimized coarse grid is usually orders of magnitude less than that with uniform fine grid.

According to the numerical analysis above, discretization induced numerical error provides additional artificial diffusion terms on each flow direction, which are determined by both numerical viscosity values and the magnitude of second order velocity derivatives. These second order derivative terms can be in different magnitudes depending on the flow characteristics, providing an opportunity of manipulating the CFD grid size along different directions without significantly increasing grid-induced error. Wang and Zhai (2012) tested this hypothesis on a few typical indoor flows with representative flow and heat transfer mechanisms.

Figure 7.16 shows one of the tested cases—forced convection (FC) driven by external forces such as mechanical force without the consideration of heat transfer (i.e., all the boundary conditions in the domain have the same temperature with no

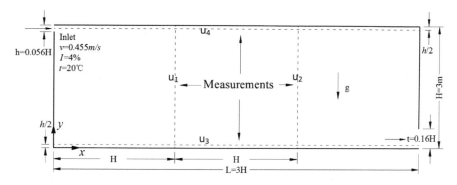

Fig. 7.16 Configuration, boundary conditions and measurements of the FC case

heat source or sink). This can be often found in built environment where the space is mainly mechanically or wind ventilated with negligible temperature difference.

Table 7.7 compares the grid resolution, the computing time, and the differences between prediction and measurement for CFD simulations with both the grid-independent and optimized coarse grids. Figure 7.17 shows the grid distribution of the optimized coarse grid. Local refinement is specified in the normal direction of inlet and outlet flow as well as wall boundaries. Other than that, the grid size is around 1/10 of the length scale of geometry (i.e., the height of the space). Normalized RMSE

Table 7.7 Comparison of grid resolution, computing time and simulation results for the FC case

Grid index	Grid number (X × Y)	Computing cost (%)	Normalized RMSE compare to experimental data			
			U1	U2	U3	U4
Grid-independent	300*100	t = 100	0.2096	0.1181	0.4510	0.2033
Optimized coarse	30*32	t = 5.8	0.2088	0.1149	0.3890	0.1583

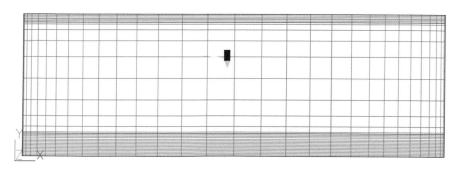

Fig. 7.17 Distribution of optimized coarse grid for the FC case

is defined in Chap. 9 that quantitatively compares the difference between two sets of data. The comparison indicates that with only 5.8% of the grid-independent CFD computing cost, the optimized coarse grid produces an almost identical prediction as the grid-independent one. This is further confirmed in Fig. 7.18, which presents the profile plots of CFD results against the experimental data.

The calculated turbulence viscosity of the grid independent solution of the FC case, with the RNG k-ε model turbulence model, is shown in Fig. 7.19. The maximum value is at the magnitude of 0.01. To evaluate the effect of grid size on the prediction accuracy, Fig. 7.20 shows the contour of predicted numerical viscosity for both fine and coarse grids in the flow domain using the same color range. It reveals that the grid independent solution has a negligible value that "totally" eliminates the numerical viscosity, but the coarse grid has a numerical viscosity comparable to (or greater) than turbulent viscosity. It needs to point out that even the grid independent one has v_x slightly greater than turbulence viscosity near the ceiling adjacent to the inlet. Ideally, the grid needs to be further refined to achieve grid independency. But for most of the computational region, the effect of numerical viscosity has been completely eliminated (Wang et al. 2014).

For the coarse grid, still v_x near horizontal walls and v_y near vertical walls are much greater than turbulent viscosity. However, the second order derivative terms, as plotted in Fig. 7.21, when near top and bottom walls, have:

$$\frac{\partial^2 u}{\partial x^2} \ll \frac{\partial^2 u}{\partial y^2} \qquad (7.48)$$

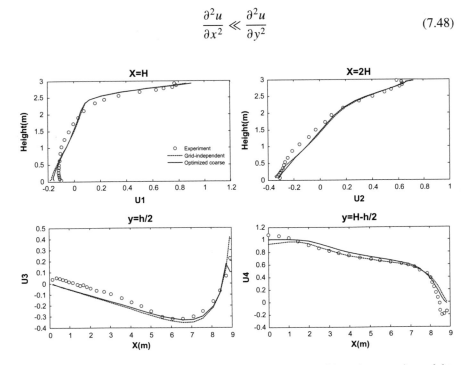

Fig. 7.18 Profile comparisons of predictions with fine and coarse grids against experimental data for the FC Case

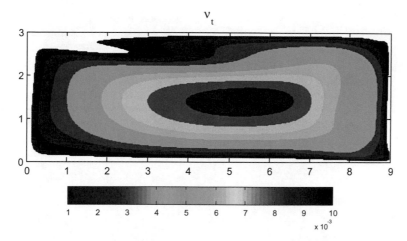

Fig. 7.19 Turbulent viscosity in the grid-independent solution for the FC Case

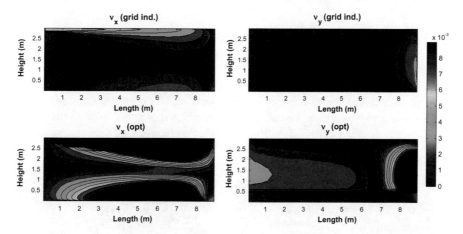

Fig. 7.20 Numerical viscosity in the grid independent (top) and the optimized coarse grid (bottom) solutions for the FC Case

This allows a large flexibility to v_x, or in other words, to the grid size in the X direction. Hence, a coarse grid can be utilized for these areas; but in the Y direction, the grid should be fine enough to eliminate the error. Similarly, when near two vertical walls:

$$\frac{\partial^2 u}{\partial x^2} \gg \frac{\partial^2 u}{\partial y^2} \tag{7.49}$$

This allows a large flexibility to v_y, or in other words, to the grid size in the Y direction. A coarse grid can be applied for these regions; but in the X direction, the

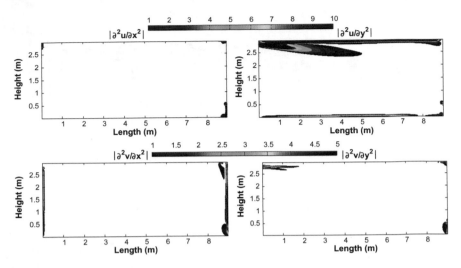

Fig. 7.21 Calculated second order derivative terms for the FC case (U-top, V-bottom)

grid should be refined to eliminate the error. This grid arrangement can ensure the total grid-induced error for grid independent and coarse grid solutions to be close to each other, as depicted in Fig. 7.22.

In sum, grid-induced error (e.g., artificial/numerical diffusion), is not only determined by the magnitude of numerical viscosity but also the corresponding derivatives that multiply the numerical viscosity. This provides the opportunity to use coarse grids with optimized cell distribution in CFD. Practically, CFD users can specify

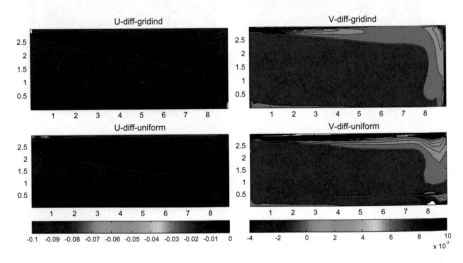

Fig. 7.22 Total diffusion values in the grid independent grid (top) and optimized coarse grid (bottom) solutions for the FC Case

a coarse grid to predict the general flow pattern of a model, comparing the orders of magnitude of diffusion and determining the fine and coarse grid areas in the domain. The final report of ASHRAE research project 1418-TRP (Zhai and Wang 2011) concludes, through a large number of tests on different typical indoor cases, that large gradient (second-order derivative value) exists in the normal direction of flow inlet/outlet and heat source/sink, as well as flow attaching to a wall; a local refinement is thus required in these key areas. A maximum of 10% geometry scale is suggested for the grid size for other areas to capture the general flow pattern. As a rule of thumb, assuming the characteristic length of a geometry under investigation is L, grid size of $1\%L$ is recommended for the local refinement areas, while for other areas and directions, $10\%L$ is suggested. The goal of using a coarse grid is mainly to reduce the computing cost. The geometry height is usually a good representative of characteristic length as indicated by different test cases. With at most around 5% of the original computing cost, the optimized coarse grid according to these guidelines can have comparable numerical results as grid independent solutions.

Practice-7: 3-D Indoor Modeling with Optimized Coarse Grids

Example Project: Buoyancy-Driven Natural Ventilation in a Confined Space
Background:
A buoyancy-driven natural ventilation room (with one large opening—door) was tested as a benchmark for CFD validation by Jiang and Chen (2003). The detail configuration of the experiment is shown in Fig. 7.23. The test chamber adjacent to the environmental chamber had a heater inside. The chamber system was located inside a larger room. The chambers were well insulated, and the wall was assumed to be adiabatic. Air velocity and temperature were measured along five vertical poles, marked P1–P5.
Simulation Details:
The test chamber was modelled in CFD by applying the developed coarse grid rules, demonstrating the application in a more realistic indoor flow scenario. Only the test chamber was simulated, with poles P2–P5 used for validation of the simulation.

According to the general rules of coarse grid specification, the local refinement was applied to heat source/sink location for this buoyancy-driven natural ventilation case. Figure 7.24 shows the grid distribution of the optimized coarse grid from the view of two vertical intersections. Because of the radiation influence from the heater, all inner surfaces act as a heat source to the space, and thus, local refinement was needed for these surfaces. The measured and predicted temperatures are normalized by the exhaust air temperature and the surrounding (ambient) air temperature as

$$T = \frac{t - 25}{33 - 25} \tag{7.50}$$

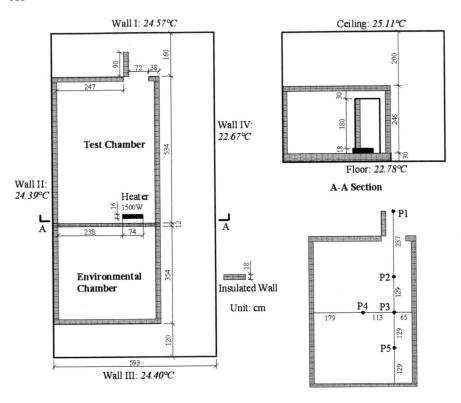

Fig. 7.23 Configuration of test chamber with buoyancy-driven natural ventilation

Table 7.8 summarizes the two grids tested, as well as the relative computing costs and the predicted normalized RMSE values as compared to the measurement. The RNG k-ε turbulence model was used, along with the standard boundary condition settings for the heater, opening (door), and walls/ceiling/floor.

Results and Analysis:

Table 7.8 verifies that with about 5% of the original computing cost, the coarse grid produces as a comparable prediction as the fine grid (with the similar normalized RMSE values). Further quantitative comparisons of predicted velocity and temperature at P2–P5 are presented in Fig. 7.25. This detailed comparison confirms the closeness of the predictions with the fine and coarse grid. The disparities between simulation and measurement may be attributed to a variety of factors in both modeling and experiment, such as, turbulence model, radiation model, boundary conditions, and measurement uncertainties.

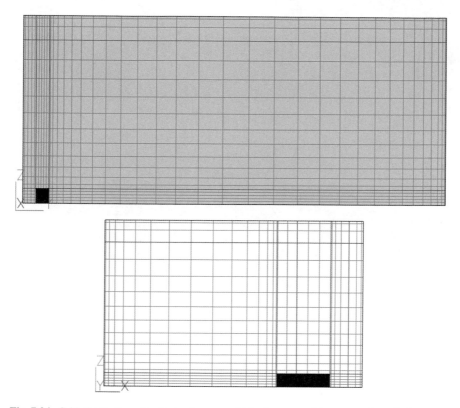

Fig. 7.24 Grid distribution of optimized coarse grid for buoyancy-driven natural ventilation case

Table 7.8 Grid resolutions, computing costs and normalized RMSE results

Grid index	Grid number (X*Y*Z)	Computing cost (%)	Normalized RMSE to experimental data	
			V	T
Grid independent	80*78*50	t = 100	0.3989	0.1400
Optimized coarse	33*40*19	t = 5.0	0.4033	0.2043

Assignment-7: Simulating Mixed Convection in a Confined Space

Objectives:
This assignment will use a computational fluid dynamics (CFD) program to model the combined-mechanical-buoyancy-driven mixed convection in a confined 2-D space.

   Key learning point:

- Importance of local grid refinement

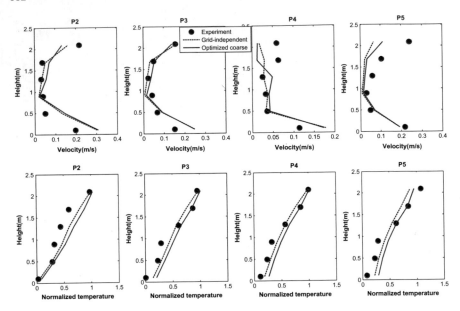

Fig. 7.25 Profile comparison of different grid solutions against experimental data

- Quasi-dynamic simulation

Simulation Steps:

(1) Build a confined space with given dimensions as shown in Fig. 7.26;
(2) Prescribe proper boundary conditions including inlet, outlet, and four walls;
(3) Select a turbulence model: the RNG k-ε model (or similar);
(4) Define convergence criterion: 0.1%;
(5) Set iteration: at least 2000 steps for steady simulation or 100 steps for each transient step;

Cases to Be Simulated:

(1) Generate and test a coarse grid using the rules in Sect. 7.5;
(2) Generate and test a fine grid with proper local refinement: at least 500,000 cells;
(3) Simulate the case with quasi-dynamic approach (unsteady simulation starting from calm indoor condition with uniform indoor temperature at 15 °C) using the coarse grid from (1).

Report:

(1) Case descriptions: description of the cases
(2) Simulation details: computational domain, grid cells, convergence status

- Figure of the grids used (on X-Y plane);
- Figure of simulation convergence records.

(3) Result and analysis

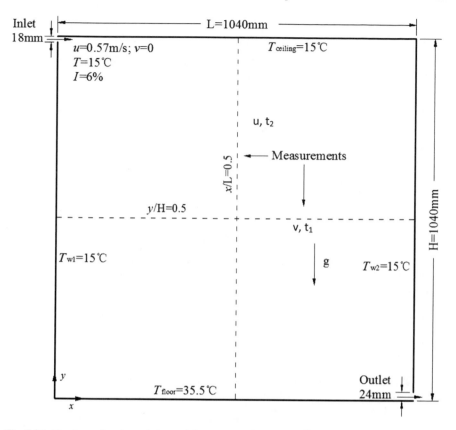

Fig. 7.26 Configuration, boundary condition setup and measurements of the MC case

- Figure of flow vectors;
- Figure of pressure contours;
- Figure of velocity contours;
- Figure of temperature contours;
- Evaluate the influences of grid refinement on simulation;
- Present the transient development of indoor conditions in unsteady simulation;
- Compare the final-state results from both steady and quasi-dynamic simulations;
- Validate the simulations with experimental data (Blay et al. 1992).

(4) Conclusions (findings, result implications, CFD experience and lessons, etc.)

References

Blay D, Mergui S, Niculae C (1992) Confined turbulent mixed convection in the presence of a horizontal buoyant wall jet. Fundam Mix Convect, HTD 213:65–72

Brown JW and Churchill RV (2014) Complex variables and applications. The 9th Edition, McGraw Hill Education. ISBN-13: 978-0073383170, ISBN-10: 0073383171

Cevheri M, McSherry R, Stoesser T (2016). A local mesh refinement approach for large—Eddy simulations of turbulent flows, 82(5):261–285

Eiseman PR, Smith RE (1990) Applications of algebraic grid generation. In: AGARD, applications of mesh generation to complex 3-D configurations, 12 p (SEE N90-21975 15-34)

Jiang Y, Chen Q (2003) Buoyancy-driven single-sided natural ventilation in buildings with large openings. Int J Heat Mass Transf 46(6):973–988

Nordanger K et al (2015) Simulation of airflow past A 2D NACA0015 airfoil using an isogeometric incompressible navier-stokes solver with the spalart-allmaras turbulence model. Comput Methods Appl Mech Eng 290:183–208

Pola FPB, Pola IRV (2019) Optimizing computational high-order schemes in finite volume simulations using unstructured mesh and topological data structures. Appl Math Comput 342:1–17

Rhie CM, Chow WL (1983) Numerical study of the turbulent flow past an airfoil with trailing edge separation. AIAA J 21:1525–1532

Römer U, Schöps S, De Gersem H (2017) A defect corrected finite element approach for the accurate evaluation of magnetic fields on unstructured grids. J Comput Phys 335:688–699

Thomas P, Middlecoff J (1980) Direct control of the grid point distribution in meshes generated by elliptic equations. AIAA J 18(6):652–656

Wackers J et al (2017) Can adaptive grid refinement produce grid-independent solutions for incompressible flows? J Comput Phys 344:364–380

Wang H, Zhai Z (2012) Application of coarse grid CFD on indoor environment modeling: optimizing the trade-off between grid resolution and simulation accuracy. HVAC&R Res 18(5):915–933

Wang H, Zhai Z, Liu X (2014) Feasibility of utilizing numerical viscosity from coarse grid CFD for fast turbulence modeling of indoor environments. Build Simul: Int J 7(2):155–164

Zhai Z, Wang H (2011) Optimizing the trade off between grid resolution and simulation accuracy: coarse grid CFD modeling, Final Report for Project 1418-TRP to ASHRAE, 159 p

Chapter 8
Solve Case

8.1 General Solution Methods

(1) Final Form of Discretized Governing Equations

Chapter 5 presents the final universal form of the discretized governing equations for all flow conservations (e.g., mass, momentum, energy):

$$A_P \phi_P = \sum_{nb} A_{nb} \phi_{nb} + S_U, \quad nb = W, E, S, N, B, T \tag{8.1}$$

where

$$A_P = \sum_{nb} A_{nb} - S_P \tag{8.2}$$

$$S_U = S_U' + S_U^T \tag{8.3}$$

$$S_P = S_P' - S_P^T \tag{8.4}$$

The main coefficients A_{nb} that relate the principal unknown ϕ_P to its neighbors ϕ_{nb} contain the combined contribution from convection and diffusion. Equation (8.1) represents a set of algebraic equations describing the conservative features of the flow on each discrete cell. Many algorithms are available to solve this set of algebraic equations. As an example, OpenFOAM (one popular open-source CFD code) implements the following methods to solve the linear algebraic problem:

- GAMG (Geometric-Algebraic Multi-Grid) for both symmetric and asymmetric matrices;
- PBiCG (Preconditioned Biconjugate Gradient) for asymmetric matrices;
- PCG (Preconditioned Conjugate Gradient) for symmetric matrices;
- smoothSolver (solver using a smoother for both symmetric and asymmetric matrices);

© Springer Nature Singapore Pte Ltd. 2020
Z. Zhai, *Computational Fluid Dynamics for Built
and Natural Environments*, https://doi.org/10.1007/978-981-32-9820-0_8

- ICCG (Incomplete Cholesky preconditioned PCG solver, i.e. PBiCG with DIC);
- BICCG (Diagonal Incomplete LU preconditioned PBiCG solver, i.e. PCG with DILU).

(2) **Direct Methods**

- *Direct Matrix Method*

  Equation (8.1) can be reformatted into

$$A_P\phi_P - A_W\phi_W - A_E\phi_E - A_S\phi_S - A_N\phi_N - A_B\phi_B - A_T\phi_T = S_U \qquad (8.5)$$

where ϕ_P is the variable (e.g., U, V, W, T, P, C) to be solved at the present point, and ϕ_{nb} is the variable at the immediate neighboring points in three directions (W, E is in x direction; S, N in y direction; and B, T in z direction). S_U is the source term that may include the influences from non-adjacent neighbors such as WW, EE, SS, NN, BB, TT, depending on the numerical schemes used. Equation (8.5) works for all the discrete points in the domain. Since all the variables ϕ in the domain must be solved simultaneously due to the high nonlinearity of the flow. Equation (8.5) can be rewritten into the following matrix format:

$$\mathbf{A} \cdot \mathbf{X} = \mathbf{B} \qquad (8.6)$$

where:

$$\mathbf{A} = \begin{pmatrix} A_{11} & \cdots & A_{1N} \\ \vdots & \ddots & \vdots \\ A_{N1} & \cdots & A_{NN} \end{pmatrix} \qquad (8.7)$$

$$\mathbf{X} = \begin{pmatrix} \phi_1 \\ \vdots \\ \phi_N \end{pmatrix} \qquad (8.8)$$

$$\mathbf{B} = \begin{pmatrix} S_1 \\ \vdots \\ S_N \end{pmatrix} \qquad (8.9)$$

$\mathbf{X}$ is the matrix for the variables to be solved in the domain, which can be extremely large for a 3D simulation with fine grid. Note that the matrix $\mathbf{A}$ is a sparse matrix, which means most of the coefficients are zero except those for the direct adjacent points to the current point to be solved as shown in Eq. (8.5). $\mathbf{X}$ can be calculated directly in theory as below:

$$\mathbf{X} = \mathbf{A}^{-1} \cdot \mathbf{B} \qquad (8.10)$$

A^{-1} is the inverse of the matrix A. To obtain the A^{-1} not only requires significant amount of computing time for large matrix but also may encounter singular issue due to the sparse nature of the matrix A (i.e., the determinant $|A| = 0$).

Example 1

$$\begin{pmatrix} 9 & 7 & 0 \\ 3 & 12 & -4 \\ 0 & -6 & 8 \end{pmatrix} \begin{pmatrix} x_1 \\ x_2 \\ x_3 \end{pmatrix} = \begin{pmatrix} 8 \\ 6 \\ 15 \end{pmatrix} \tag{8.11}$$

$$\begin{pmatrix} x_1 \\ x_2 \\ x_3 \end{pmatrix} = \begin{pmatrix} 9 & 7 & 0 \\ 3 & 12 & -4 \\ 0 & -6 & 8 \end{pmatrix}^{-1} \begin{pmatrix} 8 \\ 6 \\ 15 \end{pmatrix} = \begin{pmatrix} 3/20 & -7/60 & -7/120 \\ -1/20 & 3/20 & 3/40 \\ -3/80 & 9/80 & 29/160 \end{pmatrix} \begin{pmatrix} 8 \\ 6 \\ 15 \end{pmatrix}$$

$$= \begin{pmatrix} -3/8 \\ 13/8 \\ 99/32 \end{pmatrix} \tag{8.12}$$

- *Tri-Diagonal Matrix Algorithm (TDMA)*

Gaussian Elimination is a very useful and efficient technique for directly solving systems of algebraic equations, particularly for special cases of tridiagonal systems. The method may not be as fast as some others. Approximately, N multiplication are required in solving N equations. In addition, round-off errors may be accumulated through such many algebraic operations when N is large. The main idea of the method is to transform the system to an upper triangle array by eliminating some of unknowns.

The tridiagonal matrix algorithm, also known as the Thomas algorithm (named after Llewellyn Thomas), is a simplified form of Gaussian Elimination that can be used to solve tridiagonal systems of equations. A tridiagonal system may be written as

$$A_i \phi_i = B_i \phi_{i+1} + C_i \phi_{i-1} + D_i \tag{8.13}$$

where $A = A_P$, $B = A_E$, $C = A_W$, $D = S_U + A_N \phi_N + A_S \phi_S + A_T \phi_T + A_B \phi_B$.

In matrix form, this equation system is written as

$$\begin{pmatrix} A_1 & -B_1 & 0 & \ldots & 0 \\ -C_2 & A_2 & -B_2 & \ldots & 0 \\ 0 & -C_3 & A_3 & -B_3 & 0 \\ 0 & 0 & \ldots & \ldots & \ldots \\ 0 & 0 & 0 & -C_n & A_n \end{pmatrix} \begin{pmatrix} \phi_1 \\ \phi_2 \\ \phi_3 \\ \vdots \\ \phi_n \end{pmatrix} = \begin{pmatrix} D_1 \\ D_2 \\ D_3 \\ \vdots \\ D_n \end{pmatrix} \tag{8.14}$$

The TDMA method computes the correct values in the whole array ϕ by defining two auxiliary arrays, P and Q, as below:

$$\phi_i = P_i\phi_{i+1} + Q_i \tag{8.15}$$

or

$$\phi_{i-1} = P_{i-1}\phi_i + Q_{i-1} \tag{8.16}$$

Inserting Eq. (8.16) into Eq. (8.13) yields:

$$A_i\phi_i = B_i\phi_{i+1} + C_i(P_{i-1}\phi_i + Q_{i-1}) + D_i \tag{8.17}$$

This removes the dependence of ϕ_i on ϕ_{i-1}. Solving Eq. (8.17) with respect to ϕ_i gives:

$$\phi_i = \frac{B_i}{A_i - C_i P_{i-1}}\phi_{i+1} + \frac{C_i Q_{i-1} + D_i}{A_i - C_i P_{i-1}} \tag{8.18}$$

Comparing Eq. (8.18) with Eq. (8.15), they are the same if P_i and Q_i are computed as:

$$P_i = \frac{B_i}{A_i - C_i P_{i-1}} \tag{8.19}$$

$$Q_i = \frac{C_i Q_{i-1} + D_i}{A_i - C_i P_{i-1}} \tag{8.20}$$

The TDMA method first uses the forward process with Eqs. (8.19) and (8.20) to compute the values of **P** and **Q** arrays, from $i = 1$ to $i = n$. It then starts the backward process at $i = n$ to calculate the values of ϕ array by using Eq. (8.15). Note that in theory the algorithm is only applicable to matrices that are diagonally dominant, which is to say

$$|A_i| > |B_i| + |C_i| \quad i = 1, 2, \ldots, n \tag{8.21}$$

TDMA is a one-dimensional solution procedure. When it is used for three-dimensional problems, each row of cells in a structured grid is treated as one-dimensional system. The contributions from the neighboring points not on this row, with the most recent values, are included in the source term as shown in Eq. (8.13). The solution process is conducted for all the rows on this dimension and then repeated for the other two dimensions. Iteration is needed to reach a convergent solution.

Example 2 Calculate the temperature distribution in a solid cylinder of 3 meters, assuming $A_i = 10$, $B_i = 4$, $C_i = 5$, $D_i = 50$ for all i, and the distance between each T is 1 m (Fig. 8.1).

$$T_1 \qquad T_2 \qquad T_3 \qquad T_4$$

Fig. 8.1 1-D conductive heat transfer in a solid cylinder

(1) *If* $T_1 = 10\,°C$; $T_4 = 100\,°C$
   Forward process:

$$T_1 = 10\,°C = P_1 T_2 + Q_1 \Rightarrow P_1 = 0;\ Q_1 = 10$$
$$P_2 = B_2/(A_2 - C_2 P_1) = 4/(10 - 5 \times 0) = 0.4;$$
$$Q_2 = (C_2 Q_1 + D_2)/(A_2 - C_2 P_1) = (5 \times 10 + 50)/(10 - 4 \times 0) = 10$$
$$P_3 = B_3/(A_3 - C_3 P_2) = 4/(10 - 5 \times 0.4) = 0.5;$$
$$Q_3 = (C_3 Q_2 + D_3)/(A_3 - C_3 P_2) = (5 \times 10 + 50)/(10 - 5 \times 0.4) = 12.5$$

Backward process:

$$T_3 = P_3 T_4 + Q_3 = 0.5 \times 100 + 12.5 = 62.5\,°C$$
$$T_2 = P_2 T_3 + Q_2 = 0.4 \times 62.5 + 10 = 35\,°C$$

(2) *If* $dT/dx = 10$ *at* T_1, $T_4 = 100\,°C$

Forward process:

$$dT/dx = 10 = (T_2 - T_1)/dx \Rightarrow T_1 = T_2 - 10 \times dx = P_1 T_2$$
$$+ Q_1 \Rightarrow P_1 = 1;\ Q_1 = -10 \times dx = -10$$
$$P_2 = B_2/(A_2 - C_2 P_1) = 4/(10 - 5 \times 1) = 0.8;$$
$$Q_2 = (C_2 Q_1 + D_2)/(A_2 - C_2 P_1) = [5 \times (-10) + 50]/(10 - 5 \times 1) = 0$$
$$P_3 = B_3/(A_3 - C_3 P_2) = 4/(10 - 5 \times 0.8) = 2/3;$$
$$Q_3 = (C_3 Q_2 + D_3)/(A_3 - C_3 P_2) = (5 \times 0 + 50)/(10 - 5 \times 0.8) = 25/3$$

Backward process:

$$T_3 = P_3 T_4 + Q_3 = 2/3 \times 100 + 25/3 = 75\,°C$$
$$T_2 = P_2 T_3 + Q_2 = 0.8 \times 75 + 0 = 60\,°C.$$

(3) **Iterative Methods**

Iterative methods are commonly used in CFD due to nonlinearity of implicitly formulated governing equations and the large number of variables to be solved (i.e., values at discrete grid nodes in the domain). Iterative methods calculate the solution by iteratively updating the intermediate solution until the final one is converged. Various iterative schemes are available in the numerical analysis. This section only introduces one of the most efficient and useful point-iterative procedure for large system of equations—the Gauss-Seidel method.

- *Gauss-Seidel Method*

The Gauss-Seidel method is extremely simple but converges under certain conditions related to "diagonal dominance" of the coefficient matrix. Fortunately, the differencing of many steady-state conservation statements provides this diagonal dominance. The method takes advantage of the sparseness of matrix coefficients. The simplicity of the procedure is demonstrated below:

- Make initial guess of all unknowns (except one);
- Solve each equation for the unknowns whose coefficients are largest in magnitude using guessed values initially, and the most recently computed values thereafter;
- Repeat iteratively until changes in unknowns become small ($X = X_{calculated} + X_{error}$).

Expanding Eq. (8.6) yields:

$$A_{11}\phi_1 + A_{12}\phi_2 + \cdots + A_{1N}\phi_N = S_1 \tag{8.22}$$

$$A_{21}\phi_1 + A_{22}\phi_2 + \cdots + A_{2N}\phi_N = S_2 \tag{8.23}$$

$$\cdots$$

$$A_{N1}\phi_1 + A_{N2}\phi_2 + \cdots + A_{NN}\phi_N = S_N \tag{8.24}$$

If the diagonal elements are non-zero, rewriting each equation into:

$$\phi_1 = [S_1 - (A_{12}\phi_2 + \cdots + A_{1N}\phi_N)]/A_{11} \tag{8.25}$$

$$\phi_2 = [S_2 - (A_{21}\phi_1 + \cdots + A_{2N}\phi_N)]/A_{22} \tag{8.26}$$

$$\cdots$$

$$\phi_N = [S_N - (A_{N1}\phi_1 + \cdots + A_{NN-1}\phi_{N-1})]/A_{NN} \tag{8.27}$$

The general expression for any row ϕ_i is:

$$\phi_i = \left[S_i - \sum_{j=1, j\neq i}^{N} A_{ij}\phi_j \right]/A_{ii} \quad (i = 1, 2, \ldots, N) \tag{8.28}$$

Solve Eq. (8.28) one-by-one with the most recent updated values (using the initial guessed values if no updated ones are available), and repeat the calculations till a convergent ϕ is achieved. A convergence is reached when the absolute value of the relative approximate error $|\eta_i|$ is less than a prespecified tolerance for all unknowns (typically 0.1%).

$$|\eta_i| = \left| \frac{\phi_i^{new} - \phi_i^{old}}{\phi_i^{new}} \right| \times 100\% \tag{8.29}$$

A sufficient, but not necessary, condition for convergence of the GS procedure is that **A** is diagonally dominant:

$$|A_{ii}| \geq \sum_{j=1, j \neq i}^{N} |A_{ij}| \quad for \ all \ i \ \text{and} \ |A_{ii}| > \sum_{j=1, j \neq i}^{N} |A_{ij}| \quad for \ at \ least \ one \ i$$

$$\tag{8.30}$$

Note that if a system of linear equations is not diagonally dominant, check to see if rearranging the sequence of the equations may form a diagonally dominant matrix. In terms of computing effort, for a general system of equations, the multiplications per iteration could be as great as N; but could be much less if matrix is sparse.

Example 3 Solve Eq. (8.11) in Example 8.1.

$$9x_1 + 7x_2 = 8 \leftrightarrow x_1 = (8 - 7x_2)/9$$
$$3x_1 + 12x_2 - 4x_3 = 6 \leftrightarrow x_2 = (6 - 3x_1 + 4x_3)/12$$
$$-6x_2 + 8x_3 = 15 \leftrightarrow x_3 = (15 + 6x_2)/8$$

Table 8.1 shows the Gauss-Seidel iteration results for Example 8.3, starting from the initial values of zero for all the variables. It takes 20 iterations to reach the convergent results, which are the same as in Example 8.1.

8.2 Velocity-Pressure Decoupling Algorithms

(1) Classification of Pressure–Velocity Decoupling Algorithms

In solving governing equations of fluid by CFD, the coupling characteristics of velocity and pressure in the momentum equations pose a major challenge. There is no explicit equation to solve pressure in these governing equations. Various pressure-velocity decoupling algorithms have been proposed since the invention of CFD method and some were successfully applied in different engineering problems.

There are two main types of methods to solve the discretized algebraic equations of the momentum equations: the coupled method and the segregated method (Fig. 8.2). The coupled method is characterized by simultaneous solution of velocity and pressure. Due to the low computational efficiency and large memory requirement, it has not been widely used in general engineering applications and is mostly applied for aerospace and multiphase flow simulations. The coupled methods have been usually

Table 8.1 The Gauss-Seidel iteration results for Example 8.3

Iteration	X1	X2	X3
0	0	0	0
1	0.888889	0.277778	2.083333
2	0.67284	1.026235	2.644676
3	0.090706	1.358882	2.894162
4	−0.16802	1.506725	3.005044
5	−0.28301	1.572433	3.054325
6	−0.33411	1.601637	3.076228
7	−0.35683	1.614616	3.085962
8	−0.36692	1.620385	3.090289
9	−0.37141	1.622949	3.092212
10	−0.3734	1.624088	3.093066
11	−0.37429	1.624595	3.093446
12	−0.37468	1.62482	3.093615
13	−0.37486	1.62492	3.09369
14	−0.37494	1.624964	3.093723
15	−0.37497	1.624984	3.093738
16	−0.37499	1.624993	3.093745
17	−0.37499	1.624997	3.093748
18	−0.375	1.624999	3.093749
19	−0.375	1.624999	3.09375
20	−0.375	1.625	3.09375

employed for the computation of compressible flows, whereas the segregated methods have been preferred for incompressible flows. Different from the coupled method, the segregated method shows certain advantages of reduced computer memory and CPU time requirements (Hauke et al. 2005), computational efficiency (Haroutunian et al. 1993), and suitability for incompressible fluids (Benim and Zinser 1986) that are common in built and natural environments.

The segregated methods can be further categorized into: the pressure Poisson equation method, the artificial compression method, the pressure correction method, and the penalty method. The pressure Poisson equation method converts the momentum equations to the Poisson equations by combining the continuity equation with the momentum equations for divergence. The artificial compression method transfers incompressible fluid equations into compressible fluids, in which a pseudo-time term is added to the momentum equations, and the artificial compression term is added to the continuity equation, so the momentum equations and the continuity equation are recoupled that velocity and pressure can be solved successively. The artificial compression method requires relatively smaller time step, thus limits its application for time-dependent problems. The penalty method (Temam 1968) was named through a negative constant penalty parameter used to multiply on pressure parameter for

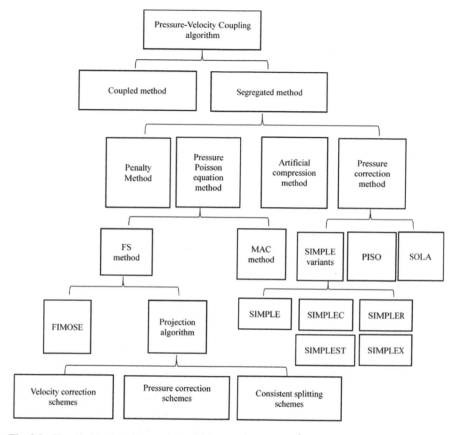

Fig. 8.2 Classification of different pressure–velocity decoupling algorithms (Wang et al. 2018)

velocity divergence. Its applicability on incompressible flow was verified (Zaparoli EL 2011). The penalty method has been used in different aspects of engineering problems, such as, for rotary machine analysis (Pelletier et al. 1991), fluid-solid interaction analysis (Kerh et al. 1998; Vincent et al. 2007), and non-Newtonian fluid flow analysis (Huang et al. 1999).

The pressure correction methods have many sub-categories such as the Maker and Cell (MAC) method, the Fractional Step (FS) method, the SOLA (Solution Algorithm) algorithm, and the SIMPLE (Semi-Implicit Method for Pressure Linked Equations) algorithm. Most (if not all) commercial CFD software adopt the pressure corrected methods, among which the SIMPLE algorithm family is most popular. The SIMPLE algorithm proposed by Patankar and Spalding (1972) is a classic and the most used pressure-correction method in engineering CFD simulations. Despite the prevalence and outstanding applicability since its invention, there are many different algorithms proposed upon it for certain aspects of improvements, which are: the SIM-PLER (SIMPLE Revised) algorithm proposed by Patankar (1980), the SIMPLEC

(SIMPLE Consistent) algorithm proposed by Van Doormal and Raithby (1984), the SIMPLEST algorithm (SIMPLE ShorTened) proposed by Spalding (1980), and the SIMPLEX (SIMPLE eXtrapolated pressure gradients) algorithm proposed by Van Doormaal and Raithby (1985).

(2) The SIMPLE Algorithm

The SIMPLE algorithm is a semi-implicit method for solving the coupled mass and momentum equations and is a numerical method primarily for solving incompressible flow fields. The key of this algorithm is to solve the Navier-Stokes (N-S) equations with a pressure "prediction-correction" step. In the process of constructing the correction equation, the pressure-correction equation is simplified by numerical manipulations so that the pressure correction value is not affected by the velocities of adjacent nodes. The pressure correction equation constructed is thus the Poisson equation. The following presents the principles of the SIMPLE algorithm.

By separating the influence of pressure term on the momentum from the source term, the discretized momentum Eq. (8.1) for U can be rewritten as:

$$A_P U_P = \sum_{nb} A_{nb} U_{nb} + \frac{P_w - P_e}{\Delta x} \Delta V + S_U, \quad nb = W, E, S, N, B, T \quad (8.31)$$

Reorganizing Eq. (8.31) provides

$$U_P = \frac{\sum_{nb} A_{nb} U_{nb} + S_U}{A_P} + \frac{A}{A_P}(P_w - P_e) = h_P + D_P(P_w - P_e) \quad (8.32)$$

$A = \Delta V/\Delta x$ is the surface area of w and e (assuming the same). Under a guessed pressure field P^*, the solution of Eq. (8.32) gives

$$U_P^* = h_P^* + D_P\left(P_w^* - P_e^*\right) \quad (8.33)$$

The values of P^* at the cell-faces are calculated from linear interpolation between two adjacent cell-centers lying on either side of the faces. In general, the velocity U_P^* will not satisfy the continuity equation due to the guessed pressure field. Subtracting Eq. (8.33) from Eq. (8.32) and neglecting the term ($h_P - h_P^*$) lead to the velocity correction at the cell-center

$$U_P = U_P^* + D_P\left(P_w' - P_e'\right) \quad (8.34)$$

$$P_w = P_w^* + P_w' \quad (8.35)$$

$$P_e = P_e^* + P_e' \quad (8.36)$$

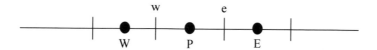

Fig. 8.3 One-dimensional illustration of control volume centers (W, P, E) and surfaces (w, e)

The same process can be applied for the discretized momentum Eq. (8.1) for V, which yields:

$$V_P = V_P^* + E_P\left(P_s' - P_n'\right) \tag{8.37}$$

$$P_s = P_s^* + P_s' \tag{8.38}$$

$$P_n = P_n^* + P_n' \tag{8.39}$$

Indeed, the same process can be applied to obtain the velocities at the w-face and e-face if considering w and e as the center of the control volume while W, P, and E are the control volume surfaces (Fig. 8.3).

$$U_w = U_w^* + F_w\left(P_W' - P_P'\right) \tag{8.40}$$

$$U_e = U_e^* + F_e\left(P_P' - P_E'\right) \tag{8.41}$$

The velocities at the s-face and e-face, respectively, are

$$V_s = V_s^* + G_s\left(P_S' - P_P'\right) \tag{8.42}$$

$$V_n = V_n^* + G_n\left(P_P' - P_N'\right) \tag{8.43}$$

By using these cell-face velocity expressions to the integral form of the continuity equation (using 2-D for example), i.e.

$$A_e U_e - A_w U_w + A_n V_n - A_s V_s = 0 \tag{8.44}$$

the continuity equation is converted to a pressure correction equation:

$$A_e U_e^* + A_e F_e\left(P_P' - P_E'\right) - A_w U_w^* - A_w F_w\left(P_W' - P_P'\right) + A_n V_n^* \\ + A_n G_n\left(P_P' - P_N'\right) - A_s V_s^* - A_s G_s\left(P_S' - P_P'\right) = 0 \tag{8.45}$$

$$(A_e F_e + A_w F_w + A_n G_n + A_s G_s)P_P' = A_e F_e P_E' + A_w F_w P_W' + A_n G_n P_N' \\ + A_s G_s P_S' - S_P \tag{8.46}$$

$$S_p = A_e U_e^* - A_w U_w^* + A_n V_n^* - A_s V_s^* \tag{8.47}$$

A_e, A_w, A_n, A_s are the surface areas. S_p is the mass conservation residual at each control volume during the iteration, which approaches zero when a convergent velocity field is reached. Once the P correction (P′) is attained via Eq. (8.46), the previous P can be updated:

$$P_P = P_P^* + P_P' \tag{8.48}$$

The velocities at the surfaces (w, e, s, n) can then be updated using Eqs. (8.40)–(8.43).

The SIMPLE algorithm was originally developed for staggered grids that store pressure information at the cell center (e.g., P, W, S, B) while storing the velocity information to the faces (e.g., w, s, b) of the cell. Under this circumstance, the two adjacent pressure nodes directly appear in the discretized momentum equation, becoming the driving force of the flow. The simulation based on a staggered grid system is robust and effective in removing the problems associated with the pressure term and the continuity equation. However, this kind of grid system needs two sets of data arrays to store the results and introduces extra complexity to the computation.

If a collocated (non-staggered) grid system is used in the simulation (which is common for most today's CFD programs), where all the variables are stored at the discrete computation nodes (e.g., the central points of cells), the linear interpolation method to obtain the pressures at cell faces (w, e, s, n, b, t) usually leads to non-physical oscillation or the so-called red-black checkerboard splitting of the pressure filed and associated difficulties in obtaining a converged solution. One widely used solution to avoid checkerboard splitting for cell-centered arrangement is to use the momentum interpolation method (MIM) (Rhie and Chow 1983) to calculate the cell face variables from the cell centered quantities.

Using the momentum interpolation procedure, the velocity at the w-face can be written as

$$U_w^* = \overline{h_{P,w}^*} + \overline{D_{P,w}}(P_W^* - P_P^*) \tag{8.49}$$

where the overbar refers to the linear interpolation of those quantities for the cells P and W. Similar to Eq. (8.34), the velocity correction at the w-face can then be obtained

$$U_w = U_w^* + \overline{D_{P,w}}(P_W' - P_P') \tag{8.50}$$

Taking these face velocities (for w, e, s, n) into Eq. (8.44) and following the same procedure in Eqs. (8.45)–(8.47) provide the P′-field, which can then be used to update the pressure and velocity values at both cell-centers and cell-faces.

The general SIMPLE algorithm takes the following steps (Van Doormaal et al. 1987) to solve the discretized governing equations (using 2-D for example):

(1) Assume a velocity distribution u^* and v^*, and a pressure field p^*; and use them to compute the coefficients and constant terms in the discrete momentum equations for the first iteration.
(2) Solve the discrete momentum equations to update the velocity field.
(3) Solve the pressure correction equation with the velocity u*, v*.
(4) Correct the pressure and velocity and various coefficients.
(5) Solve all other discretized transport equations.
(6) Determine whether the convergence is achieved; if it converges, terminate the calculation; otherwise, continue Step (2) until a convergence is reached. For unsteady simulation, the results of the calculation at this time step are used as the initial values of the iteration of the next time step.

The theoretical basis of the SIMPLE algorithm has two defects:

(1) The initial pressure field and velocity field are set separately, and the initial condition does not reflect the relationship between the pressure field and the velocity field.
(2) In the deduction of the pressure correction equation, the influences of adjacent points' velocity correction are neglected.

Although these theoretical flaws do not affect the final result of convergence, they have adverse effects on the converging process of iterative computation. Therefore, the convergence and robustness of the SIMPLE algorithm has certain room for improvement. A series of SIMPLE algorithm variants thus have been proposed to improve the computational performance.

(3) **Other SIMPLE Family Algorithms**

In the SIMPLE algorithm, the under-relaxation treatment is often adopted for the pressure correction value p', while the relaxation factor is difficult to determine. Therefore, the enhancement of the velocity field and the pressure field cannot be operated simultaneously, and the converging speed is eventually affected. The challenge may be addressed if P' is used only to modify the velocity field while the pressure field is improved by other appropriate methods—the fundamentals of the **SIMPLER algorithm** (Patankar 1980). In the SIMPLER algorithm, the initial value and update of the pressure are obtained by solving the pressure equation, and the pressure correction value obtained by the pressure correction equation is only used for updating the velocity.

The **SIMPLEC algorithm** (Van Doormal and Raithby 1984) adopts partial compensation for the influence of the neighboring-point velocity correction by changing the definition of velocity correction coefficient, so as to improve the convergence of the algorithm. SIMPLEC algorithm and SIMPLE algorithm has very similar procedure except that the velocity coefficient used in SIMPLEC allows the velocity correction equation to omit terms that are less significant than those omitted in SIMPLE; however, P' should not be under-relaxed (Yin and Chow 2003).

The **SIMPLEST algorithm** is the approach adopted by Spalding (1980) in developing commercial CFD software PHOENICS. In the case of coupling pressure and velocity, the calculation step is the same as that of SIMPLE, except that the discrete scheme of convection-diffusion term is specified in SIMPLEST algorithm. Therefore, in the SIMPLEST algorithm, the diffusion term uses linear iteration and the convection terms are treated in an explicit manner that are evaluated using the derivation from the previous iteration (Chow and Cheung 1997). The convergence speed of the iteration is slow, but it is expected that using this feature prevents the iterative process from diverging due to the coupling relationship between the convective term and the pressure. This hybrid approach improves the convergence of iterative processes to facilitate simulation of severe nonlinear problems.

Converging speed of discretized governing equation is one of the key concerns for solving the velocity-pressure coupled equation. In the **SIMPLEX algorithm** (Van Doormaal and Raithby 1985), by using extrapolation to express all pressure differences in the domain in terms of the pressure difference local to the velocity, the influence of nodal values of pressure farther from a nodal velocity is accounted for.

8.3 Solution Procedure

With the establishment of the whole set of algebraic equations for flows, the following SIMPLE-based iterative calculation sequence can be carried out to obtain the solution:

(1) Initialize all field values by reasonable guess.
(2) Solve the discrete momentum equations based on the guessed pressure field, with a proper algorithm and solver, such as, TDMA or Gauss-Seidel Method.
(3) Solve the pressure-correction equation with the same algorithm and solver to obtain the pressure-correction terms at all the cell-centers; correct the convective fluxes at the cell-faces, and the velocities and pressures at the cell-centers.
(4) Solve the discrete turbulence equations (e.g., the k and ε equations) using the same algorithm and solver, if the flow is turbulent and modeled with the RANS approach.
(5) Update the eddy viscosities if turbulent and using the RANS modeling approach.
(6) Solve the scalar transport equations using the same algorithm and solver, if required.
(7) Return to Step (2) with updated field values.

The procedure is repeated till the convergent solution is reached. For unsteady simulation, the iterative procedure should be conducted for each time step and the obtained results will be used as the initial values in Step (1) for the simulation at next time step. Figure 8.4 presents the main code structure of the simulation procedure for unsteady fluid flow. Note that the same numerical algorithm/solver can be used to solve all the governing equations that share the same convection-diffusion format as in Eq. (5.35). In programing, such a solver is often coded independent of the equations. Once a specific equation is to be solved, this solver can be called with updated coefficients **A** and **B** [as in Eq. (8.6)] for that equation. With this programing

```
        N=0
        CALL INITIALIZE(P,U,V,W,K,E,T,C)
        DO T=T0+N*dT
                DO ITER=1, ITERMAX
                        CALL SOLVER(U)
                        CALL SOLVER(V)
                        CALL SOLVER(W)
                        CALL SOLVER(P)
                        CALL CORRECT(P,U,V,W,UFACE,VFACE,WFACE)
                        IF TURBULENCE
                                CALL SOLVER(K)
                                CALL SOLVER(E)
                                CALL CORRECT(VISCOSITY)
                        ENDIF
                        CALL SOLVER(T)
                        CALL SOLVER(C)
                        CALL CHECKRESIDUAL
                        IF (RESIDUAL<RESIDUAL.SET) GOTO 100
                ENDDO
100             IF (N>=NMAX) STOP
                N=N+1
        ENDDO
```

Fig. 8.4 Main code structure for SIMPLE-based CFD simulation of unsteady fluid flows

structure, multiple algorithms/solvers can be coded in one CFD program that allows users to select according to case characteristics and simulation needs.

8.4 Convergence and Stabilization

A convergent result is considered to be attained when numerical iteration will not change the result. In physics, this implies the conservations are met for all the governing equations at both macro-scale (the entire domain) and micro-scale (the individual cells). Therefore, the conservations of mass, momentum and energy as well as turbulence must be assessed for the entire computational domain and the sum of individual cells during a computation. The assurance of the overall conservations (e.g., inlet mass = outlet mass for steady flow) does not guarantee the conservations at individual cells; likewise, the conservations at individual cells may also not ensure the overall conservations through the entire computational domain (due to the accumulated errors). Hence, checking the convergence at both scales are mandatory.

For the macro-scale, both net inflows and outflows of variables (mass, momentum and energy) should be calculated. Equation (8.51) is used to estimate the residual of the conservation, which should be less than a prescribed value (e.g., 0.1%) to ensure the convergence for all variables ϕ (while the mass and energy conservation are the most concerned variables).

$$\eta_\phi = \frac{|\phi_{net-in} - \phi_{net-out}|}{|\phi_{net-in}|} \times 100\% \qquad (8.51)$$

For the micro-scale, the solutions are considered to be converged when the sum of the normalized residuals for all the cells meets the prescribed conditions (e.g., less than 10^{-6} for mass and energy and 10^{-4} for all other variables). The normalized residuals are defined as:

$$R_\phi = \frac{\sum_{cells\,P} |\sum_{nb} A_{nb}\phi_{nb} + S - A_P\phi_P|}{\sum_{cells\,P} |A_P\phi_P|} \qquad (8.52)$$

where ϕ_P and ϕ_{nb} are the variable of the present and neighboring cells, respectively; A_P is the coefficient of the variable at the present cell; A_{nb} are the correlation coefficients of the variable of the neighboring cells; and S is the source term or boundary conditions.

A convergent solution may take many iterations to reach and often may experience instability issues depending on physics complexity, model sophistication, boundary condition setting, grid quality, and numerical schemes etc. In order to assist a stable computation, both under-relaxation method and false-time step method can be employed.

(1) *Under-relaxation method*

Under-relaxation method is to update a variable ϕ with a part of the old value and a part of the new value:

$$\phi_{update} = (1 - \alpha) \times \phi_{old} + \alpha \times \phi_{new} \qquad (8.53)$$

For instance, for the pressure correction Eq. (8.48),

$$P_P = (1 - \alpha)P_P^* + \alpha P_P' \qquad (8.54)$$

α is the relaxation factor: $0 < \alpha < 1$. Smaller α leads to a smaller change of the variable at each iteration, and thus a more stable but slower convergence. Finding proper α values for various ϕ requires experience and test, while 0.5 is a good starting point for most cases.

(2) *False-time step method*

To stabilize the simulation of steady flow, a false-time step may be included to model the steady flow as an unsteady case. With an implicit method as described in Chap. 5, the discretized unsteady governing equation is

$$\left(\frac{\rho\Delta V}{\Delta t} + A_P\right)\phi_P^n = \sum_{nb} A_{nb}\phi_{NB}^n + S + \left(\frac{\rho\Delta V}{\Delta t}\right)\phi_P^{n-1} \qquad (8.55)$$

When a small Δt is adopted in the simulation, $\frac{\rho \Delta V}{\Delta t}$ can be much more dominant than A_P, and the influence of $\left(\frac{\rho \Delta V}{\Delta t} \right) \phi_P^{n-1}$ can be greater than $\sum_{nb} A_{nb} \phi_{NB}^n + S$. As a result,

$$\phi_P^n = \frac{\sum_{nb} A_{nb} \phi_{NB}^n + S + \left(\frac{\rho \Delta V}{\Delta t} \right) \phi_P^{n-1}}{\left(\frac{\rho \Delta V}{\Delta t} + A_P \right)} \approx \frac{\left(\frac{\rho \Delta V}{\Delta t} \right) \phi_P^{n-1}}{\left(\frac{\rho \Delta V}{\Delta t} \right)} = \phi_P^{n-1} \qquad (8.56)$$

Adding the time term thus slows down the convergence but stabilizes the simulation process by putting two weighting factors at the left and right side of the equation. Proper Δt can assist the convergence of a complicated steady simulation without significant oscillation of intermediate modeling results. When the solution approaches the steady, $\phi_P^n = \phi_P^{n-1}$, which is independent of the time.

Practice-8: Fast CFD Modelling

Example Project: Semi-Lagrangian-based PISO method for fast and accurate indoor modelling.

Background:

The demand for fast engineering modeling has led to various means and efforts to reduce the cost of CFD techniques. Some of these efforts include: developing simplified turbulence models such as zero-equation models (Chen and Xu 1998); reforming solution algorithms for pressure-velocity decoupling such as Pressure Implicit with Splitting of Operator (PISO) (Issa 1986) and projection methods (Chorin 1967); utilizing coarse grids (Mora et al. 2003; Wang and Zhai 2012); and employing computer hardware technology such as Graphics Processing Unit (GPU) (Cohen and Molemake 2009) and parallel/multi-processor supercomputers. Although the rapid development of computer hardware provides more powerful computing capacity, it does not address the challenge fundamentally.

Fast fluid dynamics (FFD) is a method widely used in weather prediction and atmospheric flow study (Robert 1981; Staniforth and Côté 1991). It solves the Navier-Stokes (NS) equations with a time-advancement scheme and a semi-Lagrangian (SL) scheme. For instance, Foster and Metaxas (1996, 1997) implemented the projection method (Chorin 1967) to simulate the 3D motion of hot, turbulent gas using a relatively coarse grid. Stam (1999) proposed using semi-Lagrangian advection and fast Fourier transformation to speed up the computation to a real-time or faster-than-real-time level. Zuo and Chen (2009) first applied this operator splitting algorithm to 2-D indoor environment modeling, improved the sequence of operators, tested higher orders of differencing schemes, and evaluated the accuracy levels. Zuo et al. (2010, 2012) further improved the accuracy of FFD by using the finite volume method, mass conservation correction, and a hybrid interpolation scheme. Jin et al. (2012, 2013, 2015) extended FFD to the solution of three-dimensional airflow. Liu et al.

(2016) implemented FFD in OpenFOAM (2007) with unstructured mesh, enabling the practical application of the algorithm. Even though FFD significantly accelerates the computation, its accuracy is still far from satisfaction. This study attempts to combine the semi-Lagrangian scheme with a PISO solver with the goal to increase the computation speed of PISO but without losing the accuracy (Xue et al. 2016).

Simulation Details:

(1) ***Semi-Lagrangian Advection***

A fully implicit algorithm is unconditionally stable and has no Courant–Friedrichs–Lewy (CFL) restriction (Issa 1986), thus it is commonly used in CFD. However, in solving the momentum equations numerically, the advection term is fundamentally different from others because it brings significant non-linearity. Semi-Lagrangian scheme (Courant et al. 1952) shows potential for resolving the dilemma. The idea of semi-Lagrangian scheme was originated from the advection of scalar, but it can be directly applied to vector as well.

The Lagrangian method treats the continuum as a particle system. Each point in the fluid is labeled as a separate particle. From the perspective of such particles, the observed value of (i.g., density, temperature, etc.) will remain the same within the lapse of time. The semi-Lagrangian scheme follows the procedure as described in Fig. 8.5 to obtain the observed value of next time step. An existing velocity field of current time step t provides the velocity at (any) point A of the grid. To predict point A's value of next time step $t + \Delta t$, the semi-Lagrangian method traces back to point A's upstream location B $(\overrightarrow{AB} = -\vec{v} \cdot \Delta t)$ using the current velocity. This location B may not necessarily match an exact grid node. The surrounding values of current time step will be used to interpolate the value at this specific location. This value will then be kept and assigned to point A as its observed value of next time step. Since there is no CFL condition restriction, the time step and grid size used in a semi-Lagrangian scheme are usually large, which introduces large truncation error. Higher order numerical schemes can be used to improve the accuracy in the interpolation of the method.

(2) ***Semi-Lagrangian PISO Algorithm***

An algorithm integrating semi-Lagrangian advection with the PISO algorithm is proposed as follows.

Step 1: Semi-Lagrangian Advection: Velocity

Use the semi-Lagrangian advection $((x, -t))$ to obtain a first intermediate velocity field.

$$\frac{u^* - u^n}{\Delta t} = -(u^* \cdot \nabla)u^* \Rightarrow u^* = u^n[P(x, -\Delta t)] \tag{8.57}$$

Step 2: Predictor Step: Velocity

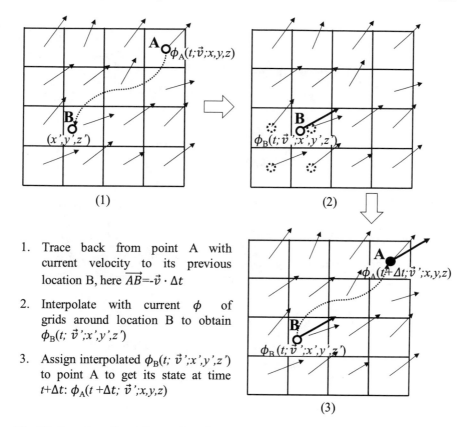

1. Trace back from point A with current velocity to its previous location B, here $\overrightarrow{AB} = -\vec{v} \cdot \Delta t$

2. Interpolate with current ϕ of grids around location B to obtain $\phi_B(t; \vec{v}';x',y',z')$

3. Assign interpolated $\phi_B(t; \vec{v}';x',y',z')$ to point A to get its state at time $t+\Delta t$: $\phi_A(t +\Delta t; \vec{v}';x,y,z)$

Fig. 8.5 Procedure of semi-Lagrangian scheme

Intermediate velocity field u^* and initial pressure field p^n are used in the solution of the implicit momentum Eq. (8.58) to yield a second intermediate velocity field u^{**}

$$\frac{u^{**} - u^*}{\Delta t} = v\nabla^2 u^{**} - \frac{1}{\rho^n}\nabla p^n + S^u \tag{8.58}$$

Since this is using p^n instead of p^{**}, ** will not satisfy the continuity equation.

Step 3: First Corrector Step: Pressure

An approximation of the velocity field u^{***} together with the corresponding new pressure field p^{***} are sought that satisfy the continuity equation

$$\nabla u^{***} = 0 \tag{8.59}$$

The momentum equation is then taken as

$$\frac{u^{***} - u^*}{\Delta t} = v\nabla^2 u^{**} - \frac{1}{\rho^n}\nabla p^{***} + S^u \tag{8.60}$$

Equation (8.60) subtracting Eq. (8.58) yields

$$\frac{u^{***} - u^{**}}{\Delta t} = -\frac{1}{\rho^n}\left(\nabla p^{***} - \nabla p^n\right) \tag{8.61}$$

Take divergence for both sides of Eq. (8.61), the velocity increment Eq. (8.61) becomes the pressure increment equation to solve $p^{***} - p^n$ field

$$\nabla^2 p^{***} - \nabla^2 p^n = \frac{\rho^n}{\Delta t}\nabla \cdot u^{**} \tag{8.62}$$

Step 4: First Corrector Step: Velocity

The updated pressure field (or pressure increment field) can be substituted into Eq. (8.60) or Eq. (8.61) to update the velocity field and produce the velocity field u^{***}.

Step 5: *Second Corrector Step: Pressure*

A replication of Step 2 is conducted using the updated result from Step 3 u^{***} and after advection of the initial value u^*, with the newest pressure field p^{****}, yields an updated velocity field

$$\frac{u^{n+1} - u^*}{\Delta t} = v\nabla^2 u^{***} - \frac{1}{\rho^n}\nabla p^{****} + S^u \tag{8.63}$$

where the explicit scheme in $v\nabla^2 u^{***}$ is taken to operate on the u^{***} field. Then u^{****} corresponding with p^{****} satisfies the continuity equation

$$\nabla u^{****} = 0 \tag{8.64}$$

Equation (8.63) subtracting Eq. (8.60) produces

$$\frac{u^{****} - u^{***}}{\Delta t} = -\frac{1}{\rho^n}\left(\nabla p^{****} - \nabla p^{***}\right) + \left(v\nabla^2 u^{***}\right) - \left(v\nabla^2 u^{**}\right) \tag{8.65}$$

After taking the divergence of both sides of Eq. (8.65), together with the continuity equation ∇u^{****} and ∇u^{***}, the velocity increment, Eq. (8.65), yields the pressure increment equation to solve the $p^{****} - p^{***}$ field:

$$\nabla^2 p^{****} - \nabla^2 p^{***} = \rho^n \nabla \cdot \left[\left(v\nabla^2 u^{***}\right) - \left(v\nabla^2 u^{**}\right)\right] \tag{8.66}$$

Step 6: Second Corrector Step: Velocity

The updated pressure field (or pressure increment field) can be plugged into Eq. (8.63) or Eq. (8.65) to update the velocity field and produce the velocity field u^{****}. More corrector steps can be used. However, the accuracy of two corrector steps is often adequate to approximate the exact solutions u^{n+1} and p^{n+1}.

The temperature field is solved separately from the velocity field, in the PISO algorithm, although the procedure is similar. Considering the coupling between the temperature and velocity, the current study used the state equation of ideal gases to update the density of air, as shown in the steps below.

Step 7: Semi-Lagrangian Advection: Temperature

Use the semi-Lagrangian advection $((x, -t))$ to obtain a first intermediate temperature field.

$$\frac{T^* - T^n}{\Delta t} = -\left(u^n \cdot \nabla\right)T^* \Rightarrow T^* = T^n[P(x, -\Delta t)] \tag{8.67}$$

Step 8: Corrector Step: Temperature

Intermediate temperature field T^* is used in the solution of the implicit energy Eq. (8.68) to yield the temperature field T^{n+1}

$$\frac{T^{n+1} - T^*}{\Delta t} = a\nabla^2 T^{n+1} + S^T \tag{8.68}$$

Step 9: Update of Density

Update the density of air with the state equation of ideal gases.

$$\rho^{n+1} = \frac{p^{n+1}M}{RT^{n+1}} \tag{8.69}$$

The proposed semi-Lagrangian PISO algorithm, without the corrector steps (Step 5 and Step 6), is similar to FFD except that it takes into consideration the pressure field from the previous time step. The FFD algorithm neglects the influence of pressure from the previous time step and assumes pressure is solely determined by the velocity field under the continuity restriction. In FFD, the advection term is completely separated from the rest of the momentum equation and is solved by using the semi-Lagrangian algorithm, which is faster and more stable compared to the conventional method of directly solving the advection equation. But the accuracy of PISO, theoretically and practically, has more advantages over FFD. The integrated algorithm (SLPISO) is expected to improve the accuracy of FFD without sacrificing much computing speed. The semi-Lagrangian advection algorithm is anticipated to largely reduce the computing cost of the direct solving of the advection term in the original PISO algorithm.

Fig. 8.6 Lid-driven cavity flow under isothermal condition

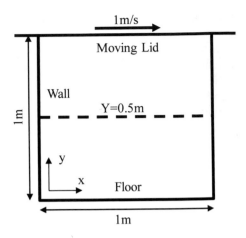

(3) *Simulation Cases*

It is critical to evaluate the performance of the developed algorithm for both steady and unsteady problems. A lid-driven cavity flow case and a mixing convection case in a confined space are used to evaluate and illustrate method performance. Figure 8.6 shows the lid-driven cavity laminar flow under isothermal condition (Ghia et al. 1982). Figure 8.7 shows a 2-D mixing convection case (Blay et al. 1992) with temperature impacts.

Results and Analysis:

(1) *Lid-driven cavity flow*

The study compares the performance of four algorithms: SIMPLE, PISO, FFD and SLPISO. The mesh size is 50×50 and the time step size is 0.005 s. Figure 8.8a

Fig. 8.7 2-D mixing convection case with heated floor

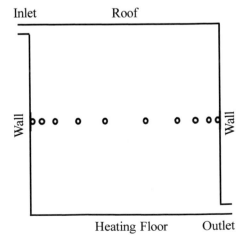

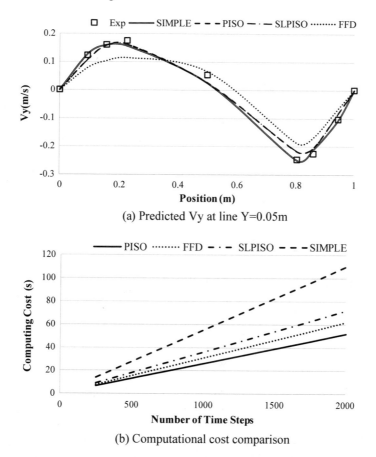

(a) Predicted Vy at line Y=0.05m

(b) Computational cost comparison

Fig. 8.8 Predicted results and computing costs of the lid-driven cavity case

shows the predicted V_y at line $Y = 0.5$ m. The results of PISO and SLPISO are almost identical. SIMPLE provides similar velocity magnitude while FFD obtains considerably different results. The results in this case reveal that SLPISO shares the same accuracy as PISO, with a slight deviation from experimental data. All of them provide better results than FFD. Figure 8.8b compares the computing costs. SIMPLE requires much more time. SLPISO has a similar speed as FFD. However, both of them are slower than PISO in this case, which will be explained later.

(2) *2-D mixing convection flow*

The study evaluates the algorithm with a 2-D mixing convection case (Blay et al. 1992) that includes the temperature field. Experimental results were obtained from the literature, which were measured in a laboratory chamber of 1.04 m × 1.04 m × 0.7 m (x × y × z) equipped with a 18 mm wide inlet slot and a 24 mm wide outlet slot. The experiment produced a fairly good 2-D flow at the central plate. The experiment measured wall temperatures and supply air conditions, respectively, as $T_{roof} = T_{walls}$ = 15 °C, T_{floor} = 35.5 °C, T_{inlet} = 15 °C, V_{inlet} = 0.57 m/s (normal to the inlet slot), as well as temperature, V_y at the ten points along the middle line on the central plate (as shown in Fig. 8.7). This study uses the constant effective kinematic viscosity and heat transfer coefficient, namely one hundred times of the physical values, to consider the turbulence impact. The mesh size is 80 × 80 and the time step size is 0.005 s. Results in Fig. 8.9 demonstrate that the SLPISO algorithm provides similar results as the PISO method. FFD has a large disparity in temperature prediction. SLPISO has similar computational speed as FFD, while they are still slower than PISO.

When the study increases the grid number from 80 × 80 to 300 × 300, and further to 1000 × 1000, the computational cost performance for these algorithms changes as shown in Fig. 8.10a, b. As the number of grid increases, SLPISO and FFD are faster than PISO. The reason for this is the inherent characteristic of the semi-Lagrangian scheme. As the grid number increases, the computing cost of the traditional solvers, such as SIMPLE and PISO, demonstrates exponential growth trend, while the semi-Lagrangian scheme shows a linear growth as revealed in Fig. 8.10c (the influence of correction steps makes the calculation cost growth of FFD and SLPISO not exactly the linear).

The comparison of simulation speed above is under the situation of using the same time step. However, the stability analysis shows that SLPISO can tolerate a larger time step than PISO. The study uses the mixing convection case with mesh size of 1000 × 1000 to check the actual calculation speed of different solvers with different time steps. Figure 8.11 shows the computing time with the largest time step that each solver can handle. To reach stable and acceptable results for this case, the largest time steps are, 0.02 s, 0.005 s, 0.08 s, and 0.1 s, for SIMPLE, PISO, SLPISO and FFD, respectively. The shadowed columns in Fig. 8.11 show the relative computing cost with the time step size of 0.005 s for all the solvers, using SIMPLE as the benchmark. The black columns show the relative computing cost using their own largest time step. While the predicted results for velocity and temperature are similar to Fig. 8.9a, b, the modeling speeds of SLPISO and FFD with larger time steps are significantly increased.

(3) *Transient 2-D mixing convection flow*

To evaluate the transient simulation accuracy of SLPISO and FFD, a transient flow in the 2-D mixing convection case is simulated. Since no transient experiment results exist for this case, the SIMPLE algorithm results are used as the reference for comparison. The "experimental" data is taken every five seconds from the SIMPLE prediction at the middle point of the test chamber. The study uses the mesh size of 80 × 80. The time step is varied from 0.005 to 0.08 s. As the time step increases, the transient results and the steady state results of SLPISO and FFD deviate, where SLPISO outperforms FFD in general (Fig. 8.12). FFD must use smaller time steps to obtain similar results as SLPISO.

The increasing deviation of the SLPISO results is attributed to the false diffusion of the time term. Compared to the original equation, the discretization of the time term

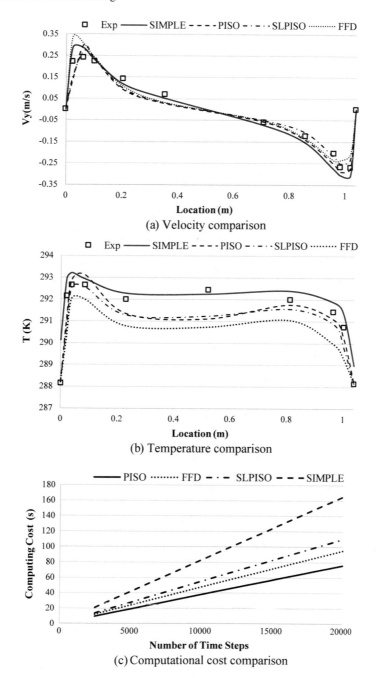

(a) Velocity comparison

(b) Temperature comparison

(c) Computational cost comparison

Fig. 8.9 Predicted results and computing costs of the mixing convection case

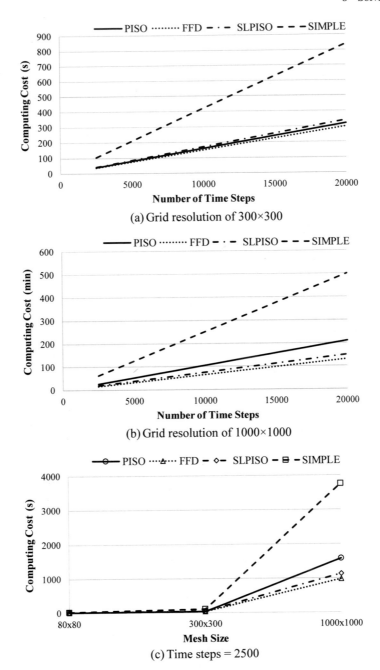

Fig. 8.10 Computational cost comparison of different solvers with different grids

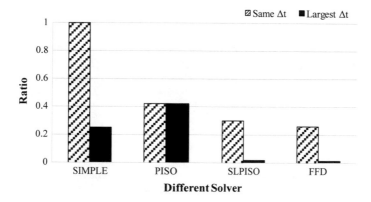

Fig. 8.11 Computational cost comparison with different time steps

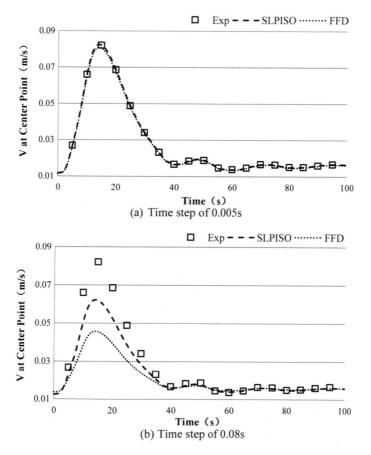

Fig. 8.12 Accuracy comparison with different time steps for transient simulation with SLPISO and FFD ("Exp" is from SIMPLE)

leads to an additional false diffusion $\frac{u^2 \Delta t}{2} \nabla^2 u$ that is related to the time step size. The increase of the time step enlarges the false diffusion, so that the transient simulation result is less responsive than the reference curve, to the transient velocity. If the constant effective kinematic viscosity is adjusted according lower, compensating for the larger time step used, the results of SLPISO with the time step of 0.08 s can be similar to the results with the time step of 0.005 s, as verified by the numerical tests.

(4) *Discussions*

Because of the inherent characteristics of the semi-Lagrangian scheme, SLPISO and FFD may not provide significant computing saving than the conventional CFD algorithms when the number of grid is relatively small. They gain their advantages when the number of grid is increased. Most engineering problems require more than one-million grids to reach solutions of grid-independence, and thus FFD and SLPISO show great potential of fast simulation for these applications. This potential is further enhanced with the advantage of being able to use larger time steps for both SLPISO and FFD. SLPISO is slightly slower than FFD but with a higher accuracy especially for transient cases. SLPISO can adopt larger time steps than PISO and FFD to obtain accurate steady state results.

Assignment-8: Simulating Microenvironment Around Thermal Manikin

Objectives:

This assignment will use a computational fluid dynamics (CFD) program to simulate the benchmark case of a computer-simulated person (CSP) under a mixing ventilation condition (Fig. 8.13).

   Key learning points:

- Indoor airflow and heat transfer simulation
- Simplification of indoor object (person)
- Comparison of simulation with experimental data.

Case Descriptions:

(1)   3D computational domain with dimensions of X × Y × Z = 2.44 × 1.2 × 2.46 m.

(2)   Air is supplied through the full cross-sectional area at one end of the channel and leaves through two circular openings at the opposite end.

(3)   The circular exhaust openings have a diameter of 0.25 m and are located 0.6 m from the floor and the ceiling, respectively.

(4)   The CSPs are located 0.7 m from the inlet, centered on the x-axis.

(5)   The geometry of the CSP is based on an average-sized woman with a standing height of 1.7 m. When seated, the CSP has a height of 1.38 m. The surface area of the CSP is 1.52 m². *Pick up a reasonable body size.*

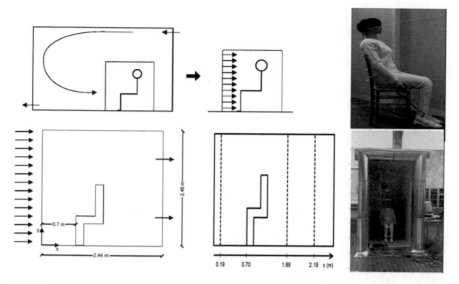

Fig. 8.13 Test of mixing convection around a thermal manikin

(6) A uniform velocity profile of U = 0.2 m/s and T = 22 °C is applied to the opening. Inlet turbulence intensity k and dissipation rate ε values can be calculated based on the literature (ISSN 1395-7953 R0307).
(7) A convective heat flow rate of 38.0 W is prescribed for the CSP corresponding to an activity level of approximately 1 Met (sedentary work).
(8) Steady state w/o contaminant.
(9) Other surfaces are adiabatic.
(10) More case details and *experimental data* can be found at: http://homes.civil.aau.dk/pvn/cfd-benchmarks/csp_benchmark_test/.

Simulation Details:

(1) Turbulence model: Re-Normalization Group (RNG) k − ε model (Yakhot and Orszag 1986).
(2) Convergence criterion: 0.1%.
(3) Iteration: at least 1000 steps.
(4) Grid: local refined grid with different total grid numbers.

Cases to Be Simulated:

(1) KERNG model with *at least* three different orders of grid numbers (e.g., 30 × 15 × 30, 45 × 23 × 45, 70 × 35 × 70).

Report:

(1) Case descriptions: descriptions of the cases.
(2) Simulation details: computational domain, grid cells, convergence status.

- Figure of the best grid used (on X–Z and X–Y planes);
- Figure of a typical convergence process recorded.

(3) Result and analysis (only present the best results except for the 1st item).

- Grid-independent solution: use one vertical pole at X = 1.69 m to compare and show the predicted velocity differences with different grids;
- Figure of velocity contours at the middle height of the CSP;
- Figure of airflow vectors at the middle height of the CSP;
- Figure of temperature contours at the middle height of the CSP;
- Figure of velocity contours at the central plane cross the CSP;
- Figure of airflow vectors at the central plane cross the CSP;
- Figure of temperature contours at the central plane cross the CSP;
- Comparison of velocities along the three tested vertical poles at the central plane cross the CSP (experiment-dot; simulation-solid line).

(4) Conclusions (findings, CFD experience and lessons, etc.)

References

Benim AC, Zinser W (1986) A segregated formulation of Navier-Stokes equations with finite elements. Comput Methods Appl Mech Eng 57(2):223–237

Blay D, Mergui S, Niculae C (1992) Confined turbulent mixed convection in the presence of a horizontal buoyant wall jet. Fundam Mixed Convect 213:65–72

Chen QY, Xu WR (1998) A zero-equation turbulence model for indoor airflow simulation. Energy Build 28(2):137–144

Chorin AJ (1967) A numerical method for solving incompressible viscous flow problems. J Comput Phys 2(1):12–26

Chow WK, Cheung YL (1997) Comparison of the algorithms PISO and SIMPLER for solving pressure-velocity linked equations in simulating compartmental fire. Numer Heat Transfer 31(1):87–112

Cohen J, Molemake JA (2009) Fast double precision CFD code using CUDA. In: 21st international conference on parallel computational fluid dynamics

Courant R, Isaacson E, Rees M (1952) On the solution of nonlinear hyperbolic differential equations by finite differences. Commun Pure Appl Math 5(3):243–255

Foster N, Metaxas D (1996) Realistic animation of liquids. Graph Models Image Process 58(5):471–483

Foster N, Metaxas D (1997) Modeling the motion of a hot, turbulent gas. In: Proceedings of the 24th annual conference on computer graphics and interactive techniques, ACM Press, Addison-Wesley Publishing Co

Ghia U, Ghia KN, Shin CT (1982) High-re solutions for incompressible flow using the Navier-Stokes equations and a multigrid method. J Comput Phys 48:387–411

Haroutunian V, Engelman MS, Hasbani I (1993) Segregated finite element algorithms for the numerical solution of large scale incompressible flow problems. Int J Numer Methods Fluids 17(4):323–348

Hauke G, Landaberea A, Garmendia I, Canales J (2005) A segregated method for compressible flow computation part I: isothermal compressible flows. Int J Numer Methods Fluids 47(4):183–209

Huang HC, Li ZH, Usmani AS (1999) Finite element analysis for transient non-Newtonian flow. Springer London Limited, London

Issa RI (1986) Solution of the implicitly discretised fluid flow equations by operator-splitting. J Comput Phys 62(1):40–65

Jin M, Chen Q (2015) Improvement of fast fluid dynamics with a conservative semi-Lagrangian scheme. Int J Numer Methods Heat Fluid Flow 25(1):2–18

Jin M, Zuo W, Chen Q (2012) Improvements of fast fluid dynamics for simulating airflow in buildings. Numer Heat Transfer Part B Fundam 62(6):419–438

Jin M, Zuo W, Chen Q (2013) Simulating natural ventilation in and around buildings by fast fluid dynamics. Numer Heat Transfer Part A Appl 64(4):273–289

Kerh T, Lee JJ, Wellford LC (1998) Finite element analysis of fluid motion with an oscillating structural system. Adv Eng Softw 29(7–9):717–722

Liu W, Jin M, Chen C, You R, Chen Q (2016) Implementation of a fast fluid dynamics model in OpenFOAM for simulating indoor airflow. Numer Heat Transfer Part A Appl 69(7):748–762

Mora L, Gadgil AJ, Wurtz E (2003) Comparing zonal and CFD model predictions of isothermal indoor airflows to experimental data. Indoor Air 13(2):77–85

OpenFOAM (2007) The open source CFD toolbox, http://www.opencfd.co.uk/openfoam.html

Patankar SV (1980) Numerical heat transfer and fluid flow. Hemisphere Publishing Corporation

Patankar SV, Spalding DB (1972) A calculation procedure for heat, mass and momentum transfer in three-dimensional parabolic flows. Int J Heat Mass Transfer 15(10):1787–1806

Pelletier D, Garon A, Camarero R (1991) Finite element method for computing turbulent propeller flow. AIAA J 29(1):68–75

Rhie CM, Chow WL (1983) Numerical study of the turbulent flow past an airfoil with trailing edge separation. AIAA J 21:1525–1532

Robert A (1981) A stable numerical integration scheme for the primitive meteorological equations. Atmos Ocean 19(1):35–46

Spalding DB (1980) Mathematical modelling of fluid mechanics, heat transfer and mass transfer processes, computational fluid dynamics unit report HTS/80/1. Imperial College

Stam J (1999) Stable fluids. In: Proceedings of the 26th annual conference on computer graphics and interactive techniques. ACM Press, Addison-Wesley Publishing Co

Staniforth A, Côté J (1991) Semi-Lagrangian integration schemes for atmospheric models—a review. Mon Weather Rev 119(9):2206–2223

Temam R (1968) Une méthode d'approximation de la solution des équations de navier-stokes. Bull Soc Math Fr 96:115–152

Van Doormal JP, Raithby GG (1984) Enhancements of the simple method for predicting incompressible fluid flows. Numer Heat Transfer Appl 7(7):147–163

Van Doormaal JP, Raithby GD (1985) An evaluation of the segregated approach for predicting incompressible fluid flows. In: National heat transfer conference, ASME, 85-HT-9

Van Doormaal JP, Raithby GD, McDonald BH (1987) The segregated approach to predicting viscous compressible fluid flows. J Turbomach 109(2):268–277

Vincent S, Randrianarivelo TN, Pianet G, Caltagirone JP (2007) Local penalty methods for flows interacting with moving solids at high reynolds numbers. Comput Fluids 36(5):902–913

Wang H, Wang H, Gao F, Zhou P, Zhai Z (2018) Literature review on pressure-velocity decoupling algorithms applied to built-environment CFD simulation. Build Environ 143:671–678

Wang H, Zhai Z (2012) Application of coarse grid CFD on indoor environment modeling: optimizing the trade-off between grid resolution and simulation accuracy. HVAC&R Res 18(5):915–933

Xue Y, Liu W, Zhai Z (2016) New semi-Lagrangian-based PISO method for fast and accurate indoor environment modeling. Build Environ 105:236–244

Yakhot V, Orszag SA (1986) Renormalization group analysis of turbulence. J Sci Comput 1:3–51

Yin R, Chow WK (2003) Comparison of four algorithms for solving pressure-velocity linked equations in simulating atrium fire. Int J Arch Sci 4(1):24–35

Zaparoli EL (2011) A comparative CFD analysis: penalty method (PM), pressure poisson equation (PPE) and the coupled formulation (PPE + PM). In: The 21st Brazilian congress of mechanical engineering

Zuo W, Chen Q (2009) Real-time or faster-than-real-time simulation of airflow in buildings. Indoor Air 19(1):33–44

Zuo W, Hu J, Chen Q (2010) Improvements on FFD modeling by using different numerical schemes. Numer Heat Transfer Part B Fundam 58(1):1–16

Zuo W, Jin M, Chen Q (2012) Reduction of numerical diffusion in the FFD model. Eng Appl Comput Fluid Mech 6(2):234–247

Chapter 9
Analyze Results

9.1 Result Visualization

With the nick name of "Colorful Fluid Dynamics", CFD is able to provide direct rendering of flow conditions that delivers the first impression on fluid trend, structure, characteristics and guides the design of fluid field or field-interacted objects. In fact, flow visualization (from both experiment and simulation) has become a separate but important discipline, from which the flow physics can be profoundly studied. Figure 9.1 illustrates the flow visualization results for hospital operating room, based on the results from both experiment and CFD simulation. The results display interesting but unexpected patterns that airflow leaving ceiling diffusers will quickly shrink to the center and form an accelerated drop-down flow towards to the patient on the surgical table. This unexpected shrink, rather than the anticipated vertical unidirectional flow, causes large vortices at sides, which may bring contaminants from backwalls and doctors to the patient and thus lead to surgical site infection (SSI).

CFD results can be presented in a set of standard visualization formats; most of these are implemented in commercial software while in-house CFD codes may use graphic software such as TechPlot to handle the CFD raw data. The conventional CFD visualization formats include:

(1) Two-dimension

  - Figure of contours of individual velocity components, speed, temperature, pressure, concentration, and turbulence on a flat plane (e.g., X-Y, X-Z, Y-Z, or any cut plane);
  - Figure of airflow vectors on a flat plane;
  - Figure of contours for post-calculated variables (such as PMV and PPD for indoor thermal comfort) on a flat plane.

(2) Three-dimension

  - Figure of iso-surfaces of individual velocity components, speed, temperature, pressure, concentration, and turbulence in the domain;

© Springer Nature Singapore Pte Ltd. 2020
Z. Zhai, *Computational Fluid Dynamics for Built
and Natural Environments*, https://doi.org/10.1007/978-981-32-9820-0_9

(a)

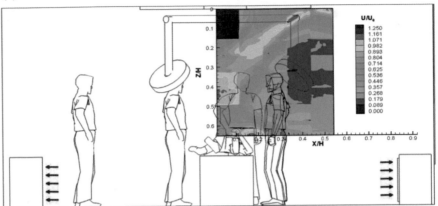

(b)

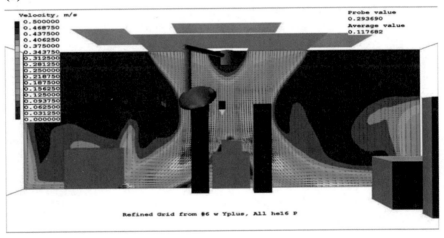

Fig. 9.1 Flow visualization results for hospital operating room airflow. **a** Measured normalized velocity contours at the central cross section of the room above patient. **b** Predicted velocity vectors and contours at the central cross section of the room

- Figure of airflow vectors in the domain;
- Figure of iso-surfaces for post-calculated variables (such as PMV and PPD for indoor thermal comfort) in the domain.

Visualization of CFD results can quickly demonstrate the flow patterns, highlight the flow features of interest, and check for the correctness of simulation results. Most unreasonable predictions can be identified from a first glance at the flow patterns such as inconsistence in contours. This can help revise the simulation before analyzing the results in a greater detail. Figure 9.2 visualizes the main flow features in the displacement ventilation case (Fig. 9.11)—a large circulation at the lower part of the space—as observed in the experiment, as well as the upward buoyancy flows above the heated objects.

9.2 Quantitative Comparison

CFD provides quantitative prediction that should be taken advantage of to obtain concrete findings. CFD programs can output spatial and temporal discrete raw data for every variable and save in a specified format (either readable or unreadable). These data can then be analyzed for further exploration and comparison. Most commercial software have implemented some basic data analysis functions, while others may count on third party data processors such as TechPlot or Excel to conduct further data analysis.

Quantitative data analysis may include various algebra calculations on CFD prediction data, such as average, root mean square etc. Comparison of key variables at particular locations of interest (e.g., where the experiment was conducted, or where the design core is located). Figures 9.3 and 9.4 present the predicted and measured velocities and temperature at nine vertical poles as illustrated in Fig. 9.12 for the displacement ventilation case (Fig. 9.11). Both the standard k-ε and 0-equation turbulence models were simulated against the experimental results. Conventionally, experimental results are presented in discrete dots as actually measured, while CFD results are presented in continuous lines because the discrete cells are artificially divided and should have no influence on the final results in physics.

9.3 Result Verification and Validation

(1) *Grid-Independent Solution*

Finding a CFD solution that is independent on grid (regardless of both grid number and distribution) is the first necessary step to evaluate the CFD prediction quality. Theoretically, when the grid size approaches zero (infinite small), the discretization introduced numerical error in CFD solution becomes zero. With a coarse mesh (large grid cell), significant numerical error is embedded in the prediction as described in Chap. 7. CFD modeling results are only valid for analysis once the grid-independency has been proved.

When analyzing CFD results, besides visualizing the variable contours and velocity vectors, predicted variables on discrete grids are usually plotted along some lines of interest, which are compared against experimental results or with those obtained from different grid resolutions. Based on the difference between two simulations with different grids, the grid convergence of the prediction can be determined. The

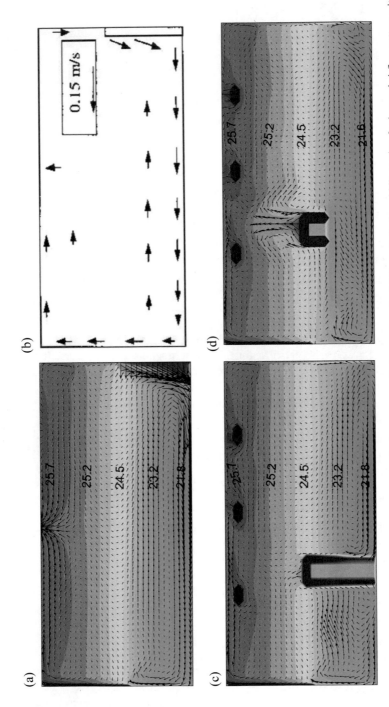

Fig. 9.2 Velocity and temperature distributions for the displacement ventilation case **a** calculated results in the middle section, **b** observed airflow pattern with smoke visualization in the middle section, **c** calculated results in the section across a computer, **c** calculated results in the section across an occupant

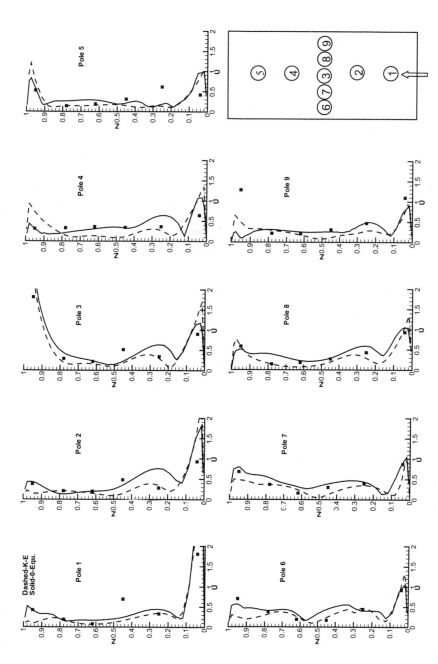

Fig. 9.3 Comparison of the velocity profiles at nine positions in the room between the calculated and measured data for the displacement ventilation case. Z = height/total room height (H), V = velocity/inlet velocity (V_{in}), H = 2.43 m, V_{in} = 0.086 m/s

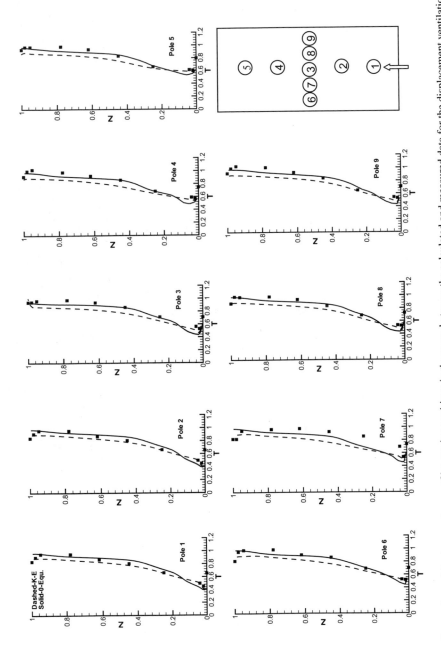

Fig. 9.4 Comparison of the temperature profiles at nine positions in the room between the calculated and measured data for the displacement ventilation case. Z = height/total room height (H), $T = (T_{air}-T_{in}/T_{out}-T_{in})$, $H = 2.43$ m, $T_{in} = 17.0$ °C, $T_{out} = 26.7$ °C

acceptability of the difference is usually judged by CFD users, mostly using their own knowledge, experience and criteria. To create a consistent and objective ground for grid independency judgment, Roache (1997) recommended using grid convergence index (GCI) as an estimator for the difference between the numerical solution and the exact solution of the governing equations:

$$GCI = \frac{3|\varepsilon|}{r^p - 1} \tag{9.1}$$

where $\varepsilon = (\phi_1 - \phi_2)/\phi_2$ is the relative prediction difference between coarse and fine grid; ϕ_1 and ϕ_2 are the prediction of coarse and fine grid respectively; $r = 2$ for grid change step (doubling or halving) and p is the order of the numerical scheme. It appears that GCI is the relative prediction difference multiplied by a constant coefficient. With a second order numerical scheme, $GCI = |\varepsilon|$.

Using GCI in practice requires proper selection of variables and locations for comparison. For example, for indoor environment modeling, the airflow and thermal environment are usually most important, which also determine the field of other variables such as species concentration. Without knowing the details of specific cases, a uniformly distributed grid in the computational domain can be recommended, which avoids the potential missing of important environmental zones. This leads to different GCI values for different locations. To evaluate the overall differences/errors from GCI results of all data points investigated, Euclidean norm (2-norm), a frequently used estimator (Celebi et al. 2010), is used to calculate the average value difference for all data points (Wang and Zhai 2012). Such an estimator is implemented previously (Wang et al. 2010) and proven to be effective for error estimation in the CFD grid independency study. The average 2-norm estimator is expressed as:

$$\frac{1}{n}\left(\sum_{i=1}^{n}|\varepsilon(i)|^2\right)^{\frac{1}{2}} = \frac{1}{n}\left(\sum_{i=1}^{n}\frac{[\phi_1(i) - \phi_2(i)]^2}{\phi_2(i)^2}\right)^{\frac{1}{2}} \tag{9.2}$$

where $\varepsilon(i) = (\phi_1(i) - \phi_2(i))/\phi_2(i)$ is the relative error on point i, and n is the number of total points under investigation. Practically, this estimator faces a problem when the variable $\phi(i)$ is a very small number that may lead to a huge $\varepsilon(i)$ number. To avoid such incidence, the average value of 2-norm of $\phi(i)$ on all point locations is thus recommended as the normalization factor so that:

$$\varepsilon(i) = \frac{\phi_1(i) - \phi_2(i)}{\frac{1}{n}\sqrt{\sum_{i=1}^{n}\phi_2(i)^2}} \tag{9.3}$$

Incorporating Eq. (9.3) into Eq. (9.2) and further Eq. (9.1) yields the new GCI index, which is similar to the conventional normalized root mean square error multiplied by a coefficient. This index is named as root mean square error (RMSE):

$$RMSE(\phi_1, \phi_2) = \frac{3}{r^P - 1} \frac{1}{n} \left(\sum_{i=1}^{n} \frac{[\phi_1(i) - \phi_2(i)]^2}{\left[\frac{1}{n} \sqrt{\sum_{i=1}^{n} \phi_2(i)^2} \right]^2} \right)^{\frac{1}{2}} = \frac{3}{r^P - 1} \sqrt{\frac{\sum_{i=1}^{n}[\phi_1(i) - \phi_2(i)]^2}{\sum_{i=1}^{n} \phi_2(i)^2}}$$

$$(9.4)$$

$RMSE(\phi_1, \phi_2)$ is derived from the GCI concept considering the comparison of predictions at different locations in a computational domain. The RMSE value provides a practically viable criterion for the CFD grid independency study. In a rigorous CFD grid independency study, the grid refinement factor r is recommended to be greater than 1.3 (Celik et al. 2008), i.e., the total grid number of a refined grid should be at least 1.3 times of the original grid.

Since different orders of numerical scheme in common use range from one to three, the average value of $p = 2$ is often presumed. If a refinement factor of 2 is employed, the coefficient on the normalized root mean square error (RMSE) criterion becomes 1. The difference between the results of two neighboring grid-resolutions thus can be computed by the following index in a uniform format:

$$RMSE(\phi_1, \phi_2) = \sqrt{\frac{\sum_{i=1}^{n}[\phi_1(i) - \phi_2(i)]^2}{\sum_{i=1}^{n} \phi_2(i)^2}}$$

$$(9.5)$$

where $\phi_1(i)$ and $\phi_2(i)$ are the predictions of the same variable at the same physical location of different grid-resolutions.

The definition of grid-independency is that the result of simulation is not affected by the density of the grid. In principle, when constantly increase the grid number, there will be a critical point where further increase in grid number will not change the prediction values. In practice, the results may still vary with the increase of grid number due to the round off error and convergent status; but there is a certain threshold in grid number where a distinct difference of RMSE is observed, which can be considered as the sign for reaching grid-independency. For practical application, the measurement uncertainty of test instrument for each compared variable (e.g., temperature, velocity, concentration) is often used as a threshold to judge the grid-independence. For instanc e, if the thermistor has a measurement error of 5%, when a RMSE less than 5% is obtained between two grids for temperature, the grid of the two with less grid number can be treated as the minimum grid number to achieve the grid-independent temperature prediction for the case in study. The same comparison should be performed for other variables in modeling. Ideally, a grid that meets all the grid-independent requirements will then be identified as the grid that can produce grid-independent solutions.

The normalized RMSE can also be used to evaluate the performance of CFD against experiment or compare two CFD simulation results with the same model. Considering the measurement error/uncertainty, the normalized RMSE value that compares the prediction with measurement is defined as:

$$RMSE(P, M) = \sqrt{\frac{\sum_{i=1}^{n} \delta_{pm}(|P(i) - M(i)| - e(i))^2}{\sum_{i=1}^{n} M(i)^2}} \qquad (9.6)$$

$$\delta_{pm} = \begin{cases} 1......if\,|P(i) - M(i)| > e(i) \\ 0......if\,|P(i) - M(i)| < e(i) \end{cases} \qquad (9.7)$$

where $P(i)$ and $M(i)$ are the prediction and measurement data sets at the same locations of interest, respectively, and $e(i)$ is the uncertainty of test instrument in the experiment. If M(i) is replaced with another simulation result Q(i), this index estimates the distance between two simulation outcomes.

It should be noted that the normalized RMSE value only yields a general impression on the performance of a prediction. This is especially true when the prediction is far away from the measurement. Detailed comparison and analysis of the predicted and measured results at critical locations is still inevitable. Combined application of normalized RMSE value and profile comparison can assist a comprehensive evaluation and analysis of prediction.

(2) *Verification and Validation of Prediction*

Chen and Srebric (2002) created a procedure for verification, validation, and reporting of indoor environment CFD analyses as the outcome of ASHRAE research project RP-1133 (Chen and Srebric 2001). This procedure, although developed for indoor environment quality study, can be used as a guideline for general CFD engineering applications. The study defines the verification and validation as follows:

- The *verification* identifies the relevant physical phenomena for flow analyses in study and provides a set of instructions on how to assess whether a particular CFD code has the capability to account for those physical phenomena.
- The *validation* provides a set of instructions on how one can demonstrate the coupled ability of a user and a CFD code to accurately conduct simulations for representative cases in study (mostly the base case of a project) with which there are experimental data available.

The American Institute of Aeronautics and Astronautics (AIAA) (1998) defines verification as "The process of determining that a (physical/mathematical) model implementation accurately represents the developer's conceptual description of the model and the solution on the model" and defines validation as "The process of determining the degree to which a (CFD) model is an accurate representation of the real world from the perspective of the intended uses of the model." The verification step is performed to ensure that a CFD code can correctly and accurately produce a solution for the mathematical equations used in the conceptual model. The verification does not imply that the computational results of a user's simulation represent the physical reality. Generally, the verification process is conducted during the CFD code development, while there is usually limited time and budget available for engineering CFD projects to conduct a systematic verification on code capabilities.

Different CFD codes have different modeling capabilities such as for fluid flow, heat transfer (conduction, convection, and radiation), mass transfer (species concentrations and solid and liquid particulates), and chemical reactions (combustion). Verification requires the use of benchmark cases that represent physical realities of interest to test the code abilities and fidelity. These cases can be simple and containing only one or more key flow and heat transfer features of the complete system. In general, the cases used for verification are not company-proprietary or restricted for security reasons. These data are often available from the literature. It is strongly recommended to report the verification. This is especially helpful in eliminating errors caused by individual users, since most existing CFD codes may have been validated by those cases. There are many examples of failed CFD simulations due to the user mistakes. Having said that, for engineering CFD projects, it is also acceptable to justify the code capabilities by referencing software manuals, example cases, and other studies on the same/similar problems using the same tool.

With the verification described above, a CFD code should be able to correctly predict the flow and heat transfer physics in the problem to be addressed. The level of accuracy depends on the criteria used in the verification. If the CFD code fails to compute correctly the flow, the problem may be:

(1) the CFD code is not capable to solve the physics (e.g., no proper governing equations);
(2) the CFD code has bugs; or
(3) there are errors in the user input data that defines the problem to be solved.

The following "Simulation Revision" Section will discuss more on possible causes for these failures.

Validation is the demonstration of the coupled ability of the user and the CFD code to accurately predict the problem of interest or the one with the same or similar physics, by comparing predictions with available experimental, analytical, empirical, or other simulation results. The fundamental strategy of validation is to identify suitable validation data, to make sure that all the important physical phenomena in the problem of interest are correctly modeled, and to quantify the error and uncertainty in the CFD simulation. Besides confirming the model capabilities and numerical accuracies, validation substantially evaluates the user's knowledge on the CFD code and his/her understanding to the basic physics involved in the flow problem as studied.

Validation confirms how accurately the user can apply the CFD code in simulating a realistic engineering problem. A CFD code may have been able to solve the abstract physical models that the user creates to describe the real world; however, the results may still not be accurate because the created models do not represent the physical reality. For example, an indoor environment may involve simultaneously conduction, convection, and radiation. A CFD user may misinterpret the problem as purely convection. The CFD prediction can be correct for the convection part; but fails in describing the complete physics involved in the actual case. It is obviously a problem at the user's side, which the validation process is also trying to eliminate.

For most cases, validation is conducted for the base case, upon which further modeling and analysis can be performed to reach meaningful conclusions. Ideally, experimental data for the tested case can be obtained from either on-site or lab measurements, with a reasonable degree of uncertainty and error. Clear descriptions on the test case with fine definition of initial and boundary conditions are highly desirable.

In practice, experimental results can be difficult to acquire. Validation may also be carried out on a similar case (to the actual problem) whose experiment is available. The validated model can then be used to alter to the actual one for simulation. As long as the two share the same fundamental physics (e.g., the same Reynolds range), no further validation is necessary for the actual case. This is particularly true and helpful for modeling design problems where the actual experimental data is not available. Such a validation still need be reported to show the fidelity of the simulation practice.

For studies with little or no experimental data, validation of simulation against analytical or empirical results is also acceptable. For instance, overall flow resistance coefficient is often applied to assess object outdoor cross-flow prediction (e.g., flow around airfoil), while ventilation efficiency can be evaluated for indoor environment study. These indices are direct outcomes of predicted variables (e.g., velocity, temperature, and concentration) and thus can provide direct evidence to support the judgment on simulation accuracy. Comparison of these *adhoc* parameters, however, only tells how accurate the modeling is; rather helps on explaining the disparities. It should be noted that it is beneficial to compare key parameters that are directly associated with simulated variables and of important interest to the study. As such, the validated model can at least predict, reasonably, the most critical (if not all) parameters of the study.

Comparing own predictions with other simulation results is the last gateway if no other data can be obtained for validation. It is important to simulate the identical case with the same geometries and boundary conditions. Different modeling methods (e.g., LES vs. RANS) and turbulence models will surely affect the comparison of predictions. Generally, results from higher accuracy simulation (e.g., DNS and LES) are more desirable to validate RANS simulation. Due to the disparities in modeling methods, model settings, grids, and inputs, exact match of two simulation results are hardly achieved. Agreement and consistence in flow trend and magnitude are valid for judging the fidelity of the simulation.

The criteria for judging verification and validation accuracy vary by cases, mostly depending on the research or application purpose. Instrument measurement error and uncertainty can be used as starting points to identify acceptable accuracy. While high accuracy is always preferred, it may not be necessary considering the high uncertainties in both test and modeling conditions (e.g., inputs). For most engineering applications, once a clear and constant trend is observed within the same magnitude order, less-than-perfect accuracy is tolerable. The validation process can be flexible, allowing a varying level of accuracy, and be tolerant of incremental improvements as time and funding permit. In principle, the validation criteria should be less restrictive for a complex case than a simple one, due to the inherent sophistication and thus uncertainties in the complex case. The criteria may also be selective. For example, if correct prediction of air velocity is more important, the criteria for heat transfer may be relaxed. Although the air velocity and temperature are interrelated, the impact of one parameter over the other may be of second order. This practice would allow the CFD user to adopt a fast and less detailed model, such as the standard k-ε model, rather than a detailed but slower model, such as low Reynolds number model for heat transfer calculation in boundary layers.

9.4   Simulation Revision

Most CFD simulations will not produce correct or reasonable prediction at the first trial. CFD prediction failure can be declared as one of the following criteria is encountered:

(1)   CFD program unexpectedly stops during simulation (produces no results);
(2)   CFD program completes the simulation but does not meet the convergence criteria;
(3)   CFD program produces converged results but results appear not correct in physics;
(4)   CFD program yields reasonable results but with large differences from experiments or others.

Many factors can contribute to the modeling failure. To verify whether the code is able to model the specific problem of interest is the first step. Besides conducting own verification as described earlier, checking software manuals, examples, and similar studies in literature using the same tool can quickly identify and exclude the issues related to the code deficiencies.

Assuming the code is capable and verified to model the physics of interest, the following provides a list of possible causes, which aligns with the general simulation process:

(a)   Model simplification, approximation and creation including domain size selection;
(b)   Problem description: buoyancy, turbulence, steady, etc.;
(c)   Physics properties specification: fluid, solid, etc.;
(d)   Boundary conditions setup: type, values, etc.;
(e)   Grid definition and generation: how much, how well;
(f)   Numerical control parameters specification: iteration, relaxation factor, convergence criteria, etc.

Failure (1) is relatively easy to fix. It is often related to Cause (a) (d) (e). Failure (2) mostly relates to Cause (a) (b) (d) (e) and (f). Failure (3) and (4) will require more efforts to diagnose the problem. Mostly they are associated with Cause (a) (b) (c) (d) and (e). "Trial and error" approach is the common method to test, diagnose and fix the challenge. In most CFD work, 10% time is spent to obtain a converged result while the rest 90% time is to ensure the results correct and reasonable. Iteration by varying various inputs (starting from self-created inputs, rather than the standard settings in the software) is inevitable for majority of CFD studies.

9.5 Parametric Study

Parametric study is one of the primary advantages of CFD. With a validated CFD model, studies can be taken to compare different modeling parameters, ranging from domain sizes, object geometries, input conditions, to turbulence models, numerical parameters. Since most of these parametric studies share the same flow mechanisms as the base case, validation is not required for these cases with alternative parameters. However, once the flow mechanisms experience fundamental shift after the change of the model parameters (e.g., increased inlet velocity changes the flow from laminar to turbulent), new validation is needed because the previous validation only works for specific flow conditions.

Most parametric studies can be carried out easily, often with one parameter changed at a time. Most of these changes are only a click in commercial software. Typically, a converged CFD model will not encounter significant convergence problem during these parametric studies unless the changes are too significant (to change the flow field largely). Results showing the correlations between altered inputs and key outputs are sought and presented in both qualitative and quantitative formats, upon which meaningful conclusions can be drawn.

Practice-9: Simulation of Building Integrated Photovoltaic-Thermal Collector

Example Project: Thermal Performance of Building Integrated Photovoltaic-Thermal Collector

Background:

The widespread adoption of photovoltaics (PV) depends heavily on reductions in installed cost per watt of generation capacity. One method for reducing cost relies on the integration of PV into building façade. These building integrated photovoltaics (BIPV) reduce installed PV cost by replacing traditional weatherproofing elements with materials that generate electricity. This approach, however, comes with an additional expense: decreased PV performance due to elevated cell temperatures and sub-optimal orientations. Decreases in installed performance may result in increased overall costs, as additional PV is required to offset the losses introduced by building integration. Davis et al. (2001) predicted that building integration may lead to cell temperatures up to 20 °C above the normal operating temperature. Since cell efficiency decreases linearly with the increase in temperature at approximately 0.4%/°C, significant reductions in cell performances can be expected in these applications.

Current approaches to PV temperature mitigation fall into two categories: natural ventilation and active heat recovery. Studies have shown that passive strategies can lower cell temperature, provide buoyancy driven natural ventilation or serve as solar air collectors for preheating HVAC supply air (Gan and Riffat 2004). Active management of PV temperatures is often coupled with systems that utilize the waste heat generated by the absorption of solar radiation that cannot be converted into electrical energy. Even with conversion efficiencies as high as 20%, up to 80% of the radiation incident on a PV panel is converted to heat, representing an enormous heating resource that can be utilized with little or no additional space requirements. Experiments have shown that the combined performance of such photovoltaic-thermal (PV/T) collectors can be as high as 70% (Athienitis et al. 2005). This study is to explore the effect of active heat recovery by a liquid cooled heat absorber on the performance of a building integrated photovoltaic-thermal (BIPV/T) collector (Corbin and Zhai 2010).

Simulation Details:

The CFD simulation solves the Reynolds Average Navier-Stokes (RANS) form of the governing partial differential equations for fluid flow and heat transfer with a finite volume method (FVM), where the RNG k-e turbulence model is employed. Because buoyancy has a critical impact on the convective heat transfer between the collector components, modeling of these effects is accomplished with the Boussinesq approximation. Radiation (the IMMERSOL model) is also included in the simulation since the radiant transfer is unique to this particular collector design.

(1) *Computer Models*

Two computer models were constructed to evaluate the performance of the BIPV/T collector under different operating conditions. A collector cooled by natural convection serves as the base case for cell temperature comparison. This model represents a standard building integrated photovoltaic installation where the PV modules are mounted close to the roof surface. The second, a collector employing a liquid-cooled tube-fin absorber into the cavity, simulates active heat recovery. The collector geometry simulated in both models matches the physical characteristics of components in the physical test array. Because each of the 5 rows in the experimental array is identical, only one row is simulated in the CFD model. The single row of PV modules is arranged in landscape format, running widthwise across the roof of the building. A schematic of the model is shown in Fig. 9.5.

The row of PV modules measures 7.3 m wide by 0.8 m deep. A roof tilt of 20 degrees above the horizontal is simulated in CFD by rotating the gravitational force 20° from vertical along the east-west axis of the row. The PV module is modeled as two sheets of glass in direct contact, each 3 mm thick, 7.3 m wide and 0.8 m deep, mounted 0.1 m above an adiabatic roof surface. The glass exposed to the environment is assigned an emissivity of 0.92. The back surface of the module is painted matte black, resulting in an emissivity of 0.95 on the surface exposed to the absorber. The silicon wafers and tedlar back sheet are not modeled as separate entities, as they are extremely thin and their resistance to heat transfer can be neglected. The upper and lower edges of the module are supported by wood blocks having an emissivity of 0.6 and measuring 0.1 m tall, 0.09 m deep and 7.3 m wide. The entire computational domain is 7.3 m wide, 0.8 m deep, 1 m tall, and includes an air space above the PV array. The adiabatic surface beneath the collector rests at the bottom of the domain. Outlets are defined above the glass blockages at the east, west, north and top boundary of the domain. Two additional outlets are defined below the glass blockages at the east and west boundaries. The south boundary of the domain is defined as an inlet.

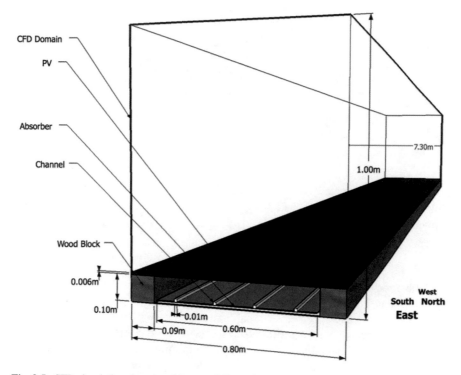

Fig. 9.5 CFD simulation domain, objects and dimensions

For the heat recovery case, additional geometry is introduced to model the tube-fin absorber. This absorber is constructed of an aluminum plate measuring 7.3 m wide by 0.6 m deep and 0.0005 m thick, which is 0.005 m above the adiabatic roof surface. Four fluid flow channels, running the full width of the absorber, are equally spaced along the top of the plate. Each channel measures 0.0085 m by 0.0085 m and is modeled as 0.0005 m thick aluminum. The entire absorber assembly is assigned an emissivity of 0.95. Each of the four flow-channels contains a fluid blockage having the properties of water. The geometry of the absorber in the simulation was simplified from the actual rhombic flow-channels and dimpled absorber plate in construction.

(2) *Boundary Conditions for Calibration Cases*

Boundary conditions used to calibrate the heat recovery model are taken from the measured ambient temperature, insolation and measured fluid inlet temperatures. The measured velocity profile is applied to the inlet at the south and the outlet at the north of the computational domain as the velocity boundary condition. This is intended to account for the increased velocity due to natural convection forces from the rows of PV modules below the row simulated. The top boundary is assigned the same velocity parallel to the inlet. Flow velocity for each fluid inlet is set to 0.345 m/s east to west and fluid density of 997 kg/m^3 is used.

Table 9.1 summarizes the temperature and insolation values used for the calibra-

Table 9.1 Parameters used for CFD model calibration

Heat recovery	T_a [C]	T_i [C]	T_s[C]	I_c [W/m^2]
Case 1	21.96	34.63	6.69	371
Case 2	23.92	34.89	9.48	498
Case 3	24.92	35.66	10.91	621
Case 4	27.19	37.17	14.16	740
Case 5	29.99	40.29	18.19	863
Case 6	33.26	44.63	22.92	991

tion cases. The equivalent sky temperature, T_s is calculated from ambient dry-bulb temperature following the equation suggested by Duffie and Beckman (1980) and applied to all air boundaries.

$$T_s = 0.0552(T_a)^{1.5} \tag{9.8}$$

Calibration is a practice that uses part of experimental results to calibrate one or few critical but uncertain input(s) by adjusting the input value(s) to match the prediction with the measurement. The calibrated input(s) will then be fixed for further studies. Predictions using the calibrated input(s) will be compared again with the other part of experiment results (not used for calibration) for validation. The initial calibration of this study showed lower than expected exiting fluid temperatures due to large heat losses at the two lower outlets at the east and west domain boundaries. In the experimental array, these outlets are highly restricted by module framing and wood blocks not modeled in the CFD simulations. By lowering the net free area ratio of these outlets, heat losses are reduced and fluid temperatures are raised to be inline with measured values. This process is repeated for each calibration case until a single net free area ratio (25%) is found that results in good agreement with measured temperatures for all cases.

(3) Boundary Conditions for Parametric Studies

The study then performs a parametric analysis for both natural convection and heat recovery models to explore the influence of environmental conditions and system designs on performance. In both natural convection and heat recovery models, east and west boundaries above and below the glass blockage are modeled as outlets with a temperature of 27 °C. A net free ration of 25% is applied to the boundaries below the glass blockage. The main domain is modeled as air, with fluid properties calculated at 20 °C. Sky temperature is calculated using the procedure discussed above for each case and applied to all air boundaries.

For the heat recovery model, the inlets at the east end of the four flow channels are each assigned a density of 997 kg/m^3, a temperature equal to the water inlet temperature being tested, and a velocity of 0.345 m/s east to west, equivalent to the measured fluid flow rate from the experimental study. All solid surfaces in the simulation have no-slip boundary condition.

Both natural convection and heat recovery cases vary isolation levels between 1000 and 250 W/m² in 250 W/m² increments. The range of insolation values are translated into total heat flux based on the following assumptions: (1) the electrical energy removed by the PV cells does not contribute to the heating of the cells; (2) the PV modules operate at 17.3% efficiency; (3) The PV glass τ · α product of 0.85 accounts for losses due to reflection at off-normal incidence angles. These assumptions result in total calculated heat fluxes of 4105, 3079, 2053 and 1026 W applied to the glass following the equation:

$$\dot{Q} = I_c A_c \tau \alpha (1 - \eta) \tag{9.9}$$

where $\dot{Q}$ is the collector heat flux, I_c is the insolation incident on the collector, A_c is the collector surface area, $\tau \cdot \alpha$ is ratio of insolation transmitted through the glass and absorbed by the cells, and η is the module electricity generation efficiency. For the active heat recovery model, the water inlet temperature in the channels is varied between 50.0 and 10.0 °C in 10.0 °C increments.

Results andAnalysis

(1) Simulation Results for Calibration Cases

The numerical study first examines the influence of CFD grid on simulation results by comparing the computed velocity values taken at 5 heights in 12 locations. The total cell numbers investigated vary between 56,640 and 251,559. The study found that the mesh with 74,160 cells can provide reasonably good grid-independent solutions and is thus selected for all subsequent simulations. The simulation time with this grid is approximately 3 h with 3000 iterations on a 2.8 GHz Pentium IV system with 512 MB of ram. The residual values for mass and energy conservation are 1.0 E-5 and 1.0 E-2, respectively.

Temperature values at the bottom surface of the lower glass blockage are obtained by exporting the predicted temperature contour into a comma separated text file. Temperatures are then read at the same locations as those measured. The bottom surface is chosen because the silicon cells composing the PV module are mounted there and it is assumed that the bottom surface temperature best approximates the silicon cell temperature. Exiting fluid temperatures for the cases simulating heat recovery are also exported and averaged for all four outlets. Simulation results are plotted in Fig. 9.6 and show reasonably good agreement with the measured data when using a net free area ratio of 25% at the two lower east and west outlets. The industry standard method of presenting solar thermal collector test and simulation data is to plot efficiency, η, versus $(T_i - T_a)/I_c$, where T_i is the collector fluid inlet temperature and T_a is the ambient dry-bulb temperature. Total heat collection efficiency, η, is the ratio of collected heat by fluid, $\dot{Q}_f = \rho C_p (T_o - T_i)$, to total insolation incident on the collector, I_c. T_o is the collector fluid outlet temperature.

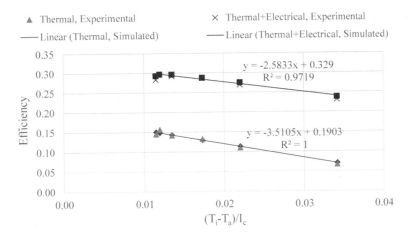

Fig. 9.6 Characteristic curves for BIPV/T collector—simulation and measured results

(2) *Simulation Results for Parametric Studies*

Air temperatures and exiting fluid temperatures are exported from the CFD model following the same process outlined in the model calibration. Figure shows the temperature contours on the bottom surface for natural convection and heat recovery cases, where the insolation is 1000 W/m² and the inlet fluid temperature is 10.0 °C. The highest temperatures are seen in the natural convection model, reaching 73.0 °C at 1000 W/m². The heat recovery model shows lower maximum temperatures of 66.9 °C and 71.0 °C at 1000 W/m² and inlet temperatures of 10.0 °C and 50.0 °C, respectively. The local influence of the absorber channels on both the absorber plate (bottom) and the PV cells (middle) (in Fig. 9.7) can be clearly seen.

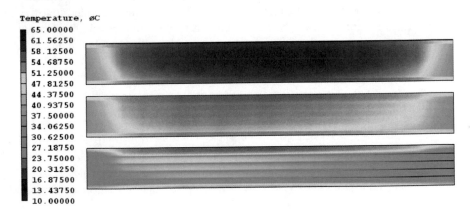

Fig. 9.7 Photovoltaic temperature contours for natural convection (top), heat recovery (middle), and thermal absorber temperature contour (bottom) at 1000 W/m² and 10 °C inlet temperature

Figure 9.8 illustrates the effect of the absorber plate on the airflow behind the PV cells. The cross section shown here is taken at the midpoint of the array, 3.65 m from the east edge. Although the absolute values of the velocities are equivalent, the pattern of the flow is much more complex around the absorber plate. The outlines of each channel are clearly visible as are the five small convection loops that form above the absorber surface. In contrast, the natural ventilation case shows one large convection loop. The formation of small convection loops is a result of increased turbulence, which increases heat transfer between the back surface and the absorber plate. Figure 9.9 compares average cell temperatures with varying insolation levels for both the natural convection and heat recovery models.

(3) Discussion of Results

CFD parametric results show that heat recovery is capable of lowering both average and maximum cell temperatures compared to natural convection. A 10 °C inlet water temperature at 1000 W/m^2 results in cell temperatures 13.2 °C lower than natural

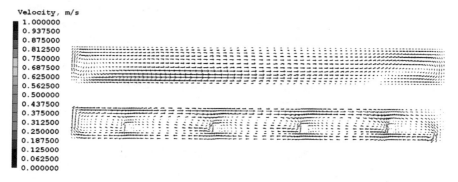

Fig. 9.8 Velocity vectors at the middle cross-section beneath photovoltaic module for natural convection (top) and heat recovery (bottom)

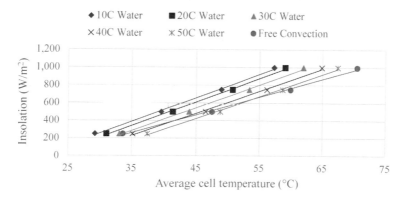

Fig. 9.9 Average cell temperatures at varying inlet temperatures and insolations—parametric study results

convection when averaged over the collector area, and 6.2 °C lower at a maximum. Water entering at 50 °C, is still capable of lowering cell temperatures over the natural convection case, but only at high insolation levels. At low insolation levels, high inlet water temperatures tend to heat the PV cells. Low inlet water temperatures have a cooling effect at all insolation levels investigated. Taking 10 °C as the water inlet temperature, a standard silicon cell with a temperature coefficient of 0.4%/°C at an average of 13.2 °C reduction in temperature results in a 5.3% increase in conversion efficiency. Those cells operating at the maximum temperature see a 2.5% increase in conversion efficiency.

Although gains in electrical output are modest for this collector, the amount of heat collected by active heat recovery can be significant. Fluid outlet temperatures, averaged from the exported temperature contours, are able to reach 51.1 °C, suitable for domestic use or for hydronic heating systems. Average temperature rise over the collector is only 1.1 °C, but collected thermal flux reaches 449.6 W at an efficiency of 11.0%. Lower inlet temperatures result in higher efficiencies and collected power. The amount of useful heat flux collected at varying insolation levels and fluid inlet temperatures is shown in Fig. 9.10. The amount of useful thermal power collected by this collector shows a clear linear relationship with insolation and inlet water temperature. Note the loss of power at high inlet water temperatures and low insolation. Also note that the amount of thermal power collected is larger than the insolation when inlet water temperatures are lower than the ambient temperature. When inlet water temperatures are higher than ambient, thermal power can be collected or lost depending on the insolation level.

Assignment-9: Simulating Displacement Ventilation in a Confined Space

Objectives:
This assignment will use a computational fluid dynamics (CFD) program to model a complex indoor environment with a 3-D side-wall supply displacement ventilation.

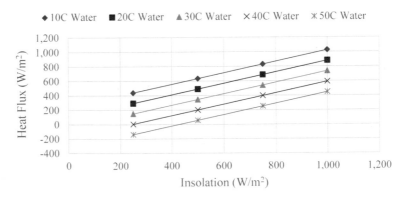

Fig. 9.10 Collected heat flux at varying inlet temperatures and insolations—parametric study results

Key learning point:

- Create complex indoor objects
- Simulate combined air, heat and contaminant flows.

Simulation Steps:

(1) Build a confined space with given dimensions as shown in Fig. 9.11;
(2) Build indoor objects as specified in Tables 9.2 and 9.3;
(3) Prescribe proper boundary conditions for all objects (including contaminants) [pay attention to the inlet conditions];
(4) Select a turbulence model: the RNG k-ε model (or similar);
(5) Define convergence criterion: 1%;
(6) Set iteration: at least 2000 steps for steady simulation;
(7) Determine proper grid resolution with local refinement: at least 500,000 cells.

Cases to Be Simulated:

(1) Simulate the steady flow case (with heat and contaminant).

Report:

(1) Case descriptions: description of the case

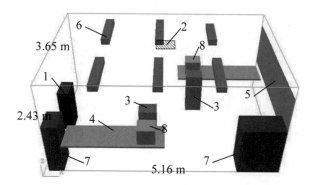

Fig. 9.11 Layout of the displacement ventilation case (inlet-1, outlet-2, person-3, table-4, window-5, fluorescent lamps-6, cabinet-7, computer-8): The room is 5.16 m × 3.65 m × 2.43 m. The displacement ventilation diffuser provides a ventilation rate of 4 ACH through the perforated front panel with a net area ratio of 10%. The equivalent air velocity through the front panel is 0.086 m/s. The supply and exhaust air temperatures are, respectively, 17 and 21.6 °C. The contaminants are released from the nose of the two persons at a rate of 40 ml/h

Table 9.2 Geometrical, thermal and flow conditions for the diffuser and window

Displacement ventilation case			
Inlet diffuser	Size: 0.53 m × 1.1 m	Temperature: 17.0 °C	Velocity: 0.086 m/s
Window	Size: 3.65 m × 1.16 m	Temperature: 27.7 °C	Closed

Table 9.3 Sizes and
capacities of the heat sources

Heat source	Size (m x m x m)	Power (W)
Person	$0.4 \times 0.35 \times 1.1$	75
Computer 1	$0.4 \times 0.4 \times 0.4$	108
Computer 2	$0.4 \times 0.4 \times 0.4$	173
Overhead lighting	$0.2 \times 1.2 \times 0.15$	34

(2) Simulation details: computational domain, grid cells, convergence status

- Figure of the grids used (on X-Y and X-Z plane and at critical regions);
- Figure of simulation convergence records.

(3) Result and analysis

- Figure of flow vectors at critical planes (e.g., across the inlet, person etc.);
- Figure of pressure contours at critical planes (e.g., across the inlet, person etc.);
- Figure of velocity contours at critical planes (e.g., across the inlet, person etc.);
- Figure of temperature contours at critical planes (e.g., across the inlet, person etc.);
- Figure of contaminant concentration contours at critical planes (e.g., across the inlet, person etc.);
- Figure of 3-D contaminant concentration iso-surfaces;
- Validate the simulation results with experimental data at P1-P9 (Fig. 9.12) (Yuan et al. 1999).

(4) Conclusions (findings, result implications, CFD experience and lessons, etc.)

Fig. 9.12 Floor plan of nine
vertical poles for velocity,
temperature and contaminant
test (P1-P9)

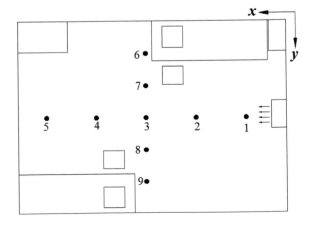

References

AIAA (1998) Guide for the verification and validation of computational fluid dynamics simulations. AIAA G-077–1998, Reston, Virginia

Athienitis A, Poissant Y, Collins M, and Liao L (2005) Experimental and numerical results for a building-integrated photovoltaics test facility. Photovoltaic specialists conference, conference record of the thirty-first IEEE, 3–7 Jan. 2005, pp 1718–1721

Celebi ME, Celiker F, Kingravi HA (2010) On euclidean norm approximations. Pattern Recogn 44(2):278–283

Celik IB, Ghia U, Roache PJ, Freitas CJ, Coleman H, Raad PE (2008) Procedure for estimation and reporting of uncertainty due to discretization in CFD applications. J Fluids Eng 130(7):078001–078004

Chen Q, Srebric J (2001) How to verify, validate, and report indoor environment modeling CFD analyses. Final Report for ASHRAE RP-1133, 58 pages

Chen Q, Srebric J (2002) A procedure for verification, validation, and reporting of indoor environment CFD analyses. HVAC&R Res 8(2):201–216

Corbin CD, Zhai Z (2010) Experimental and numerical investigation on thermal and electrical performance of a building integrated photovoltaic-thermal collector system. Energy Build 42:76–82

Davis MW, Fanney AH, Dougherty BP (2001) Prediction of building integrated photovoltaic cell temperatures. J SolEnergy Eng 123(3):200–210

Duffie JA and Beckman WA (1980) Solar engineering of thermal processes. Wiley, New York

Gan G, Riffat SB (2004) CFD modeling of air flow and thermal performance of an atrium integrated with photovoltaics. Build Environ 39(7):735–748

Roache PJ (1997) Quantification of uncertainty in computational fluid dynamics. Annu Rev Fluid Mech 29:123–160

Wang H, Zhai Z (2012) Analyzing grid independency and numerical viscosity of computational fluid dynamics for indoor environment applications. Build Environ 52:107–118

Wang B, Zhao B, Chen C (2010) A simplified methodology for the prediction of mean air velocity and particle concentration in isolation rooms with downward ventilation systems. Build Environ 45(8):1847–1853

Yuan X, Chen Q, Glicksman L, Hu S, Hu Y, Yang X (1999) ASHRAE research report: performance evaluation and development of design guidelines for displacement ventilation. ASHARE, 230 pages

Chapter 10
Write CFD Report

10.1 General Requirements for Reporting CFD Results

ASHRAE produces the Guideline 33 for Documenting Indoor Airflow and Contaminant Transport Modeling (ASHRAE 2013). This can be used as the ground for general requirements for reporting engineering CFD simulation and results. In general, four parts are required to form a complete CFD report.

(1) **General project description and study goal**
This should provide an overview of the study or project, which may include the type of problem, associated engineering issue or background, and/or actual project (location, structure, conditions, etc.). The report should explicitly clarify the goals of the modeling, such as to verify the design targets, produce empirical formula, or find appropriate technical solutions. The objectives of a simulation study can dictate the type of modeling and analysis to perform (e.g., whether radiation/contaminant to be considered) as well as the subset of modeling assumptions to be employed.

(2) **Detailed project description**
This section will provide details of the project, which should be concise but adequate for people of interest to replicate the simulation. This may involve various geometric information (dimensions for components within the computational domain), flow-related system conditions (supply, return, mixing etc.), objects that may influence the flow (obstacle, leakage, etc.), and elevation/terrain that may affect gravity and flow trend. Pictures and drawings are common materials to be included.

(3) **CFD model description**
This part should specifically indicate what CFD modeling method (e.g., RANS or LES) and what software/code are employed for the purposes of the current study. The report should include the name and version of the software, governing equations, turbulence model, convergence criteria, pressure-velocity coupling

© Springer Nature Singapore Pte Ltd. 2020
Z. Zhai, *Computational Fluid Dynamics for Built and Natural Environments*, https://doi.org/10.1007/978-981-32-9820-0_10

scheme (e.g., SIMPLE or SIMPLER), equation solution procedure (such as coupled or segregated solver), and spatial (and temporal) discretization schemes and their order of accuracy. Justifications are required for specific selections, for instance, steady state versus unsteady simulation.

Flow governing equations indicate what variables to be modeled. While the common governing equations can be briefed (e.g., the Navier-Stokes equations), special models (e.g., combustion, radiation, etc.) and treatments (e.g., thermal comfort index calculation) should be articulated. For turbulent flows, particular modeling method and turbulence model used should be clarified with justification. Uncommon turbulence model, if used, should be explained with associated constants and coefficients indicated. The justification of these selections can be based on a literature review or previous experience. Known limitations of the method and model can be listed, which may assist the analysis and explanation of obtained simulation results.

It is also important to specify the physical properties of the fluid, species, and particles, if any, to be modeled, such as density, dynamic/kinematic viscosity, heat capacity, expansion ratio, Prandtl number, Schmidt number, particle diameter and density, and thermal conductivity and diffusivity. This is particularly necessary if some unusual conditions are simulated. If radiation model is used, surface radiation properties need be provided, such as emissivity, absorptivity, and/or if a gray body assumption is used.

Clear and complete description of CFD model is most critical, which should provide the details on the computer model developed from its origin with sufficient explanations on model simplification and approximation. In most cases, both original model and final CFD model should be presented to highlight the similarity and disparities (due to simplifications and assumptions) between the two. The size of the computation domain, the sizes and locations of physical models (e.g., obstacles such as furniture, equipment, openings such as windows, doors, supply and return vents, and sources of heating, cooling, and/or contaminant species) must be illustrated with their values being tabled.

Figure 10.1 shows an example of such a CFD model illustration and description. The study was to investigate the PM2.5 concentrations caused by smoking in a typical public transportation shelter, and its public health risks arise from the

 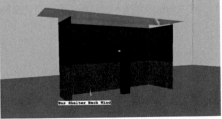

Fig. 10.1 Public transportation shelter and CFD model

second-hand smoke. The public transportation shelter in the model was designed after the real shelters commonly found in US. Dimensions of the structure are approximately 1.2 m deep, 3.6 m long, and 2.55 m high. It has three walls and a roof, with each of the walls consisting of metal mesh on the bottom half and tinted glass on the top half. To account for the mesh sections of the walls, modeling is done with 50% porous thin plates. All other walls and the roof are simply adiabatic solid plates. There are air gaps of 8 cm located between the ground and the walls, and also between the roof and the walls on the sides.

Boundary conditions differentiate cases even with the same CFD model. A complete CFD model description must include clear information on every boundary condition. Depending on the types of problems to be solved, boundary conditions can be generally divided into the following groups: airflow, thermal, contaminants, and other groups (e.g., combustion) as described in Chap. 6 Specify Boundary Conditions. The boundary conditions used or to be tested can be summarized in tables for clear presentation. The variations of the CFD model (in the physical model, or boundary conditions, or turbulence models, or numerical methods, etc.) lead to the comparison study using CFD, often called sensitivity analysis or parametric study.

The last but not least item to be included in the CFD model description is mesh. The mesh discretizes the computational domain into solvable numerical cells. The size and quality of mesh have direct impacts on the simulation correctness, accuracy and efficiency. Therefore, the mesh details must be provided including total number of meshes tested if multiple meshes are used, average size of the grids, total number of the grids, types of meshes, (e.g., structured or unstructured, uniform or nonuniform), and quality of the mesh (e.g., aspect ratio, skewness, and alignment). Typical mesh figures at represented sections are provided. If localized refinement is used, details on size and location of the regions where the refinement is applied are often provided with highlighted images. Size of the first cell from the solid surface (boundary layer) is important once turbulence model is tested or justified.

Finally, it is professional to include in the report the key simulation parameters, such as time step; total simulation time; underrelaxation factors; false-time steps, if any; maximum number of iterations; convergence criteria for each governing equation; and numerical constants for specific solvers used, such as for multigrid solvers. Information should be of the form to allow reproduction of the numerical simulation. It is also very common to provide the computer hardware employed for the study, including CPU frequency, number of CPU cores, number of computer nodes for a cluster, and total system RAM, as well as computing time per case.

(4) **CFD modeling results and analysis**

Three parts of contents are typically included in the result section.

a. **Mesh independency study**

The report should first summarize the practice on verifying solution's independency on mesh. The procedure described in Chap. 9 Analyze Results should be followed to evaluate the influences of grid number and mesh quality on simulation results. Quantitative comparisons of predicted key variables at meaningful locations with different meshes need be provided to justify whether a mesh-independent solution is achieved. For conventional modeling problems, reasonable mesh size (grid number) recommended from the literature can be used to justify the adequacy of the adopted mesh for similar simulations (in both nature and size).

b. **Validation of simulation results**
Validation is inevitable for every modeling study. Three resources may provide the information that can be used for CFD validation: (1) own experimental test results from either field test or laboratory experiment; (2) experiment results obtained from others or literature; (3) comparable simulation results from others. While detailed comparison is always desirable, lump-sum parameters such as total resistance coefficient and heat flux may also provide an overall judgement on modeling accuracy and quality. This, hence, can still be valuable if detailed test data is not available.

c. **Simulation results and analysis**
With a validated model, CFD can be used for various parametric studies within the same flow mechanism as that been validated (e.g., within the same Reynolds range). The simulation should be able to address the questions raised in the study objectives. Analysis is mandatory to discuss and explain the obtained modeling outcomes. This is particularly true when unexpected results are attained, which often requires iteration on the modeling to isolate, enhance or repeat the findings. A reasonable number of figures and tables are necessary to illustrate the main findings. These figures and tables should be self-explainable with their own captions and legends.

(5) **Conclusions and recommendations**
The report should provide a set of concrete conclusions out of the main findings obtained including lessons and experience learnt. Conclusions should indicate the degree of success to which the objectives of the study were met. Recommendations may be provided as necessary depending on the purposes of the study and should indicate how the results of the simulations support the recommendations.

(6) **Result files**
Electronic version of models and simulation results are often part of the deliverable, including immediate outputs of simulation tools as well as the secondary calculation results from the post processes that support conclusions and recommendations. In some cases, it may be prohibitive to provide all direct output files of the analysis tools (e.g., due to the file size). In such cases, the ability to reproduce the outputs using the input files should be provided.

(7) **References**
The report should provide a list of references that were used and cited to support the various phases of the study, including version information of simulation

tools, simulation tool manuals, and similar studies, as well as various referred documents in the report.

10.2 Writing CFD Reports for Technical or Course Projects

Technical or course projects often have very specific targets/goals and allow (or require) the inclusion of more details than published articles (e.g., in conference proceedings or journals). Depending on the objectives of the course or the project, the report requirements may vary from case to case. Detailed guidance on required contents (figures and tables) in the reports are usually articulated by clients (via contract) or instructors (via assignment), in order to document the accomplishment of the key tasks.

The assignments presented at the end of each chapter of this book provide examples of typical CFD course project requirements that may provide insights on a basic course project and report format upon which changes can be made according to individual interests or needs. The following sessions present one CFD technical project, as an example, to demonstrate some common report structures.

Project Title: Prediction of Duct Fitting Losses

Objectives: The primary objective of the project is to evaluate the feasibility and accuracy of using CFD techniques to numerically determine the loss coefficients for duct fittings. The success of this may eliminate the need of laboratory fitting tests in compliance with ASHRAE Standard 120 (ASHRAE 2017), and further facilitate the design process of duct systems.

Project Tasks

(1) *General Case Descriptions*:
The consultants will be asked to determine the loss coefficients for two duct fittings using CFD. Referring to Figs. 10.2 and 10.3, a flat oval straight body tee and a flat oval straight body lateral fitting need be simulated, with both converging and diverging air flows (opposite flow directions) (Sleiti et al. 2013). Figures 10.4 and 10.5 show a standard lab test apparatus in compliance with ASHRAE Standard 120.

(2) *Cases to Be Simulated*:
Total four (4) cases will be simulated by CFD, each with a variety of air flow rates.

(a) flat oval straight body tee, converging air flow;
(b) flat oval straight body tee, diverging air flow;
(c) flat oval straight body lateral, converging air flow;
(d) flat oval straight body lateral, diverging air flow.

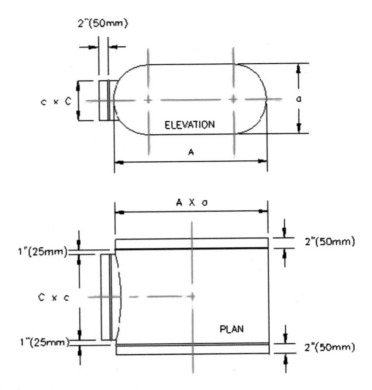

Fig. 10.2 Flat oval straight body tee

Geometry of Duct Fittings (Table 10.1):
Nominal Main Dimensions mm (in.): 787 × 356 (31 × 14);
Nominal Branch Dimensions mm (in.): 559 × 254 (22 × 10).
Range of Air Inlet Velocity and Reynolds Number:
Inlet Velocity = 1000–4000 fpm (5.1–20.3 m/s);
Common Section Reynolds Number = 85,000 to 500,000.

(3) *Simulation Details*:

- CFD software: free to choose
- Computational domain: free to choose
- Boundary conditions: free to choose
- Inflow turbulence intensity: 10%
- Surface roughness: materials: galvanized steel, $\varepsilon = 0.09$ mm (0.0003 ft.)
- Turbulence modeling method: free to choose
- Numerical scheme: free to choose
- Grid: free to choose (grid-independence of solutions must be verified)
- Convergence criterion: 0.1% for all variables (normalized by inflow conditions).

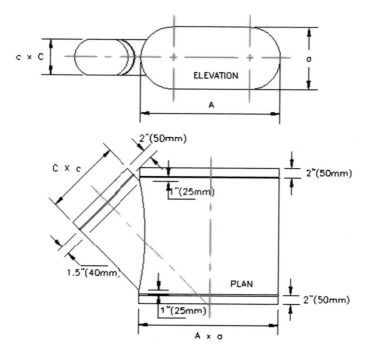

Fig. 10.3 Flat oval straight body lateral

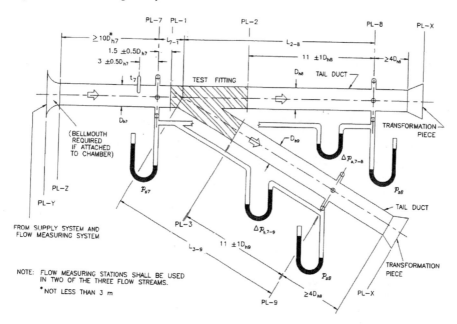

Fig. 10.4 Diverging airflow test apparatus

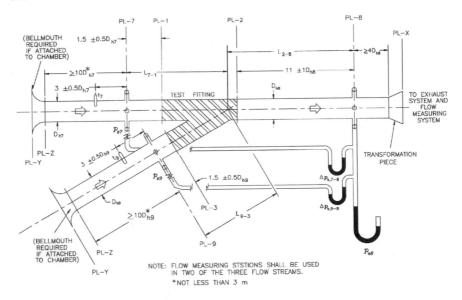

Fig. 10.5   Converging airflow test apparatus

Table 10.1   Geometry and description of duct fittings to be tested

Nominal fitting size [mm (in.)] [A × a to C × c (Fig. 10.2)]	Description
787 × 356 to 559 × 254 mm (31 × 14 to 22 × 10 in.)	Tee, straight
787 × 356 to 559 × 254 mm (31 × 14 to 22 × 10 in.)	Lateral, straight

Final Report Requirements:

(1) *Case descriptions*: descriptions of the simulated cases (including, but not limited to, geometry, boundary conditions, special treatments/assumptions, fluid and flow properties, etc.)

(2) *Simulation details*:

- Governing equations
- Turbulence modeling methods
- Numerical schemes
- CFD program algorithms and flow charts
- Numerical treatments of near wall regions (e.g., wall function) and boundary conditions
- Computational domain
- Computational grids (resolution and distribution)
- Convergence status (e.g., iteration number, convergence criteria)
- Requirements for computer hardware and software and computing time

(3) *Results and analysis*:

- Grid-independent solutions
- Verification and validation of the CFD code/user for similar cases
- Airflow characteristics in the tested duct fittings
- Calculated loss coefficient C_s (main loss coefficient) and C_b (branch loss coefficient) as a function of flow rate splits Q_s/Q_c and Q_b/Q_c [Q = flow rate]

$$C_s = \frac{\Delta p_{t,1\text{-}2}}{p_{v8}} \quad C_b = \frac{\Delta p_{t,1\text{-}3}}{p_{v9}} \tag{10.1}$$

The subscripts are: b—branch section; c—common section; s—straight (main) section; t—total; v—velocity; 1, 2, 3, 8, 9—plane number as shown in Figs. 10.4 and 10.5. The flow rate ratios should be varied over a reasonably wide and realistic range from 0.1 to 0.9.

(4) *Conclusions*.

10.3   Writing CFD Articles for Formal Publications

A CFD-based research article has almost all key elements as listed in the general reporting requirements, but with a more focused research agenda/problem to address. Limited by allowed paper length (typically less than 15 pages, single lined, and 12 font size), a research article will put more spaces for discussing the innovative research methods and/or major new findings, rather than describing well-known knowledge such as conventional governing equations and turbulence model etc. The article, however, still need to provide necessary and adequate information on these, so as to allow a complete and fair judgement on the work as well as allowing replication if interested. The following suggests a typical CFD paper structure, which may create a starting point for drafting a CFD article for formal or journal publication. A CFD conference paper may still consider this template but often has a shorter length (4–8 pages) and special formatting requirements as demanded by different conferences.

Title: A concise and precise title is the key to have the paper well received by both reviewers and readers from the first glance. It is critical to avoid a vague, ambitious, or general title. Problem/method targeted title is desirable but not a too narrowed one. A too narrowed title may question the applicability of the study for other conditions and thus the usability of the paper for most audiences. Generally, a title should not be more than 30 words.

Abstract: The abstract is the first item (and most of cases the only item) to be read by others. Therefore, it should summarize the entire paper in a concise manner, starting with problem statement and study objectives. The main research approaches (e.g., simulation or experiment) should be present. The abstract will then focus on the main contributions of this study, either in methodologies or findings. Well-known conclusions such as "CFD is a powerful tool" should be skipped, along with other findings that can be sensed before reading this article. A typical abstract has about 300–500 words, with only one paragraph. Citation and abbreviation should always be avoided unless necessary.

Keyword: 3–5 keywords are often required that represent the main topics involved in the paper. The keywords are primarily for search purpose. Well-chosen keywords can assist wide exposure and high citation of the article.

Introduction: Introduction often starts with the research background including challenges or problems to be addressed. It will be followed by a solid literature review on recent progress on the same topic of this paper. Depending on the topic and the purpose of the paper, the review may focus on the recent and major progresses including key milestones on the same issue in study. The intent is to emphasize the necessity and significance of the proposed research. Hence, besides the summary of the existing studies, critiques on current findings and/or methods are crucial. Previous conclusions can also be used to verify and/or justify the findings of this study. The introduction often ends with an explicit indication of the objectives of this paper. The typical length of introduction is about 1.5–2 pages.

Methodology: This is the portion where the applied CFD methodologies will be described, including the general CFD approach (e.g., RANS vs. LES), governing equations, turbulence models, numerical schemes, convergence criteria, special treatments and handlings, etc. Depending on whether new methods, models or treatments are introduced, the section can run from 1 page to 3–4 pages. For most studies using commercial software without adding new models/functions, this section can be brief with adequate citation to the software documents and/or other papers. Repeating conventional governing equations including classic turbulence models is not necessary unless new items or coefficients are presented.

Case Description and Validation: This section provides the details on the cases to be simulated, including both base case and case variations. Summaries of case geometries and boundary conditions are often delivered in the format of figures and tables. Assumptions and approximations should be well documented and justified. Validation includes a complete mesh independence study, followed by the comparison of the prediction with available and key measurement data. Discussion is important to explain the disparities between simulation and experiment if any. Further tests may be desired to confirm the explanations, for instance, caused by different wall treatment methods. Alternatively, conclusions from literature on the same problem can be cited to justify the disparities. A typical length of this portion is 2–3 pages.

Results and Analysis: This section presents the main simulation results, in the format of figures and tables. Most journals only allow a maximum of 15 figures and tables combined. Therefore, only those critical results should be selected. Each figure or table should present distinct and significant findings or evidence that support conclusions. These figures and tables should be self-explainable with clear captions and legends. Figures and tables are the second item to be read by most readers, after they find the paper is interesting by reading the abstract. Text should avoid simple descriptions of these figures and tables; instead, should indicate the main

Table 10.2 Evaluation criteria for ASME journals

Originality	Good
Significance	Good
Scientific relevance	Good
Completeness	Good
Acknowledgement of the Work of others by References	Good
Organization	Good
Clarity of writing	Good
Clarity of tables, graphs, and illustrations	Good
In your opinion, is the technical treatment plausible and free of technical errors?	Yes
Have you checked the equations?	Yes
Are you aware of prior publication or presentation of this work?	No
Is the work free of commercialism?	Yes
Is the title brief and descriptive?	Yes
Does the abstract clearly indicate objective, scope, and results?	Yes

findings (either as expected or unexpected). Analysis and discussion are inevitable to articulate the primary findings of the study. Comparison of the findings against those in literature is always desirable. This section can run 2–5 pages depending on the findings.

Conclusions: Conclusions is the third item to be read once one finds the figures and tables are of interest. Not being a duplication from the abstract, the conclusions should emphasize the major contributions of this paper. Detailed and explicit descriptions of the findings in either method or conclusion or both should be provided. Limitations of the study may also be briefed, along with the future research directions. A decent conclusion section is about 0.5–1 page.

References: A complete list of references that are cited by the paper should be provided in the format required by each journal. This includes all references such as theses, books, journal articles, conference papers, online resources, internal reports/handouts, as well as personal conversations.

The following Table 10.2 provides one example of review criteria from the American Society of Mechanical Engineers (ASME). Similar standards are used by most scientific journals. Checking the quality of an article against this list is a good practice for both writing and reviewing a paper.

Practice-10: Coupled Indoor and Outdoor Simulation with Air-to-Water Heat Exchange Using Multi-Block Meshes

Example Project: Investigation of cooling efficiency drop of dry-cooling towers under cross-wind conditions

Background:

Natural-draft dry-cooling tower is an energy-efficient and water-saving cooling equipment in power plants, widely used in the regions lack of water but rich in coal or oil, such as, South Africa, Middle East and North China. However, the performance of dry-cooling towers is highly sensitive to the environment conditions, particularly the wind conditions that may reduce up to 40% of the total power generation capacity (Ding 1992). The conventional design of cooling towers does not sufficiently consider the impact of wind, which in fact exists most of time in reality. Hence, it is important to investigate the influence of wind on the performance of cooling towers and propose appropriate improving measures. Figure 10.6 displays a typical Heler-type dry cooling tower with vertical heat exchangers around the bottom of the tower, which generally confronts the most significant impacts from cross-winds.

Simulation Details:

The study simulates the two full-scale cooling towers in tandem arrangement with the air–water heat exchangers vertically located at the bottom of the towers (Zhai and Fu 2006). Figure 10.7 illustrates the computational domain, the boundary conditions and the grid system used. Only half of the flow field was simulated because of the symmetry in geometry and flow/thermal conditions. A typical wind speed $U_{ref} = 10$ m/s in the winter was applied to study the impact of cross-wind on the performance of the cooling towers. The wind profile was set up as:

$$U_{wind} = U_{ref}\left(Z/Z_{ref}\right)^{0.16}, V = W = 0 \qquad (10.2)$$

where $Z_{ref} = 45$ m.

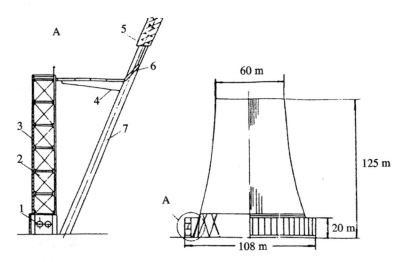

Fig. 10.6 Sketch of Heler-type dry-cooling tower: (1) water pipes, (2) heat exchangers, (3) shutter, (4) support beam, (5) shell of tower, (6) seal plate, (7) X-shape supporter

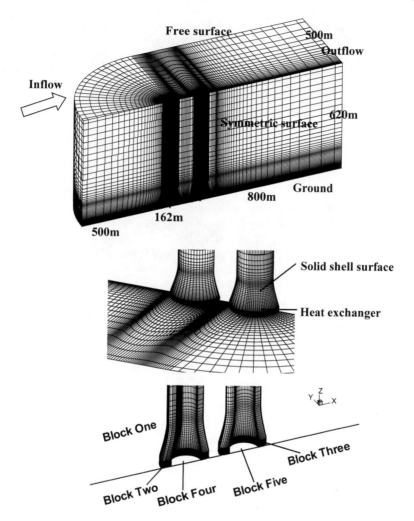

Fig. 10.7 Computational grid and five-block division for the flow over two cooling towers

The investigation uses the multi-block CFD algorithm and program developed by Zhai and Fu (2002) to simulate the airflow and heat transfer in and around two cooling towers. The program has been verified by many previous studies (Zhai 1999). The computation adopted five blocks of grids—one exterior flow block, two interior flow blocks, and two shell/heat exchanger block—to ensure the generation of high quality grids, as shown in Fig. 10.7. Each grid block was generated individually with the total grid cells of about 220,000. A fine grid with 640,000 cells was also used to verify the grid independence of numerical results.

The numerical simulation solves the steady-state governing conservation equations of mass, momentum and energy, with the standard k-ε turbulence model and wall function (Launder and Spalding, 1974) to represent the overall turbulence effect. The CFD simulation incorporates the heat exchange process between air and water of the heat exchangers, as well as the resistance effect of the heat exchangers to airflow. The heat released at the water side and absorbed at the air side of the heat exchangers are, respectively,

$$Q_{water} = C_w \dot{m}_w (T_{w1} - T_{w2}) \tag{10.3}$$

$$Q_{air} = \alpha (\overline{T_w} - T) A \tag{10.4}$$

where, C_w is the water heat capacity, $\dot{m}_w$ is the water mass flow rate in the heat exchangers, T_{w1} and T_{w2} are the water inlet and outlet temperature in the heat exchangers; α is the heat transfer coefficient of the heat exchangers and is related to the airflow velocity through the heat exchangers, $\overline{T_w}$ is the mean water temperature in the heat exchangers, T is the air temperature outside the heat exchangers, A is the surface area of the heat exchangers.

Since the air temperature and flow rate as well as water temperature are varying with locations under wind conditions, the numerical simulation divides the heat exchangers into N uniform sections around the bottom of the cooling towers and M layers in vertical direction. For the J-th section,

$$Q(J) = Q_{air}(J) = \sum_k \alpha \left[\overline{T_w(J)} - T \right] A(J)$$
$$= Q_{water}(J) = C_w \dot{m}_w(J)[T_{w1}(J) - T_{w2}(J)] \tag{10.5}$$

where Q(J) is the total heat exchange rate between air and water at the J-th section of the heat exchangers, k denotes the k-th layer in the vertical direction of the J-th section, T is the local air temperature, $\dot{m}_w(J)$, $T_{w1}(J)$ and $T_{w2}(J)$ are the water mass flow rate, water inlet and outlet temperature at the J-th section, $\overline{T_w(J)} = [T_{w1}(J) + T_{w2}(J)]/2$ is the mean water temperature in the J-th section. The heat transfer coefficient α of the heat exchangers can be obtained from the literature (Ding, 1992):

$$\alpha = 1372.34 L_2^{0.515} \left(\frac{W}{m^2 K} \right) \tag{10.6}$$

$$L_2 = C_k^{0.64} L_1 \left(\frac{ton}{m^2 h} \right) \tag{10.7}$$

where, L_1 and L_2 are, respectively, the original and modified air mass flow rate through the heat exchangers per front area, and $C_k = 1.11$.

Since the heat transfer between air and water influences the airflow velocity while the airflow velocity inversely affects the heat transfer performance, an iterative coupling algorithm is required:

(1) Assume initial fields of air velocity, temperature, turbulence, and the distribution of $T_{w2}(J)$. $T_{w1}(J)$ is specified as a constant water temperature based on power generation turbine outputs. Calculate the heat transfer coefficient α with Eqs. (10.6) and (10.7) and calculate $\overline{T_w(J)} = [T_{w1}(J) + T_{w2}(J)]/2$.

(2) Solve Eq. (10.5) to obtain $Q_{air}(J)$ and introduce this heat source term into the energy conservation equation of air.

(3) Solve the airflow governing equations to obtain the new distributions of air velocity, pressure, temperature, and turbulence.

(4) Update the heat transfer coefficient α with Eqs. (10.6) and (10.7). Calculate $Q_{air}(J)$ with previous $\overline{T_w(J)}$ values and Eq. (10.5).

(5) Calculate the new $T_{w2}(J)$ with Eq. (10.5) and update $\overline{T_w(J)}$.

(6) Go back to (2) until the solution is converged.

The total heat exchange rate $Q_{total} = \sum_J Q(J)$ and the distribution of $T_{w2}(J)$ are two major results for evaluating the cooling performance of the towers under different wind conditions and improving strategies. Q_{total} represents the overall cooling capacity or efficiency of the cooling towers, while the distribution of $T_{w2}(J)$ indicates the locations of cooling deficiency and improvement.

Heat exchangers provide not only heat sources but also resistance to air movement. The study uses the same iteration process to account for the airflow resistance effect of the heat exchangers in cooling towers. The field test shows that the air pressure drop through the heat exchangers has the following relationship to the airflow rate (Ding 1992):

$$\Delta P = 2.1L_1^{1.76} + 0.06L_1^2 (\text{Pa}) \tag{10.8}$$

L1 is the air mass flow rate through the heat exchangers per front area. Once L_1 is updated with current air velocities, the new pressure resistance term can be obtained and introduced to the momentum equation of air to produce the new velocity distribution.

Results and Analysis:

(1) *Independence of results on computational grids*

The study first examines the independence of numerical results on the grid resolution. Figure 10.8 shows the distribution of water outlet temperature in the air–water heat exchangers around the towers, predicted with two different grids. The difference of two solutions is negligible, indicating good grid-independence of the simulation. Table 10.3 further verifies that the difference of the total heat exchange rates in windy days calculated with the fine and coarse grid is only about 3%.

(2) *Comparison of numerical results with experimental tests*

To verify the creditability of numerical results, the study compares the simulation with the model experiment and available field tests (Ding 1992). Table 10.4 presents the predicted total heat exchange rate without cross-wind and with the same operating conditions as those in the design and field test. The calculated result is between the design value and the field test result, and the difference

Fig. 10.8 Water outlet
temperature in the heat
exchangers at the bottom of
towers (−80: windward; 0:
lateral; +80 leeward; U_{wind}
= 10 m/s)

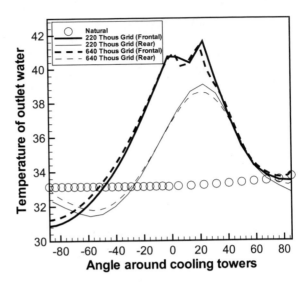

Table 10.3 Comparison of computed total heat exchange rates of cooling towers with different
simulation conditions

	Fine grid			Coarse grid	
	$U_{\text{wind}} =$ 0 m/s	$U_{\text{wind}} =$ 10 m/s windward tower	$U_{\text{wind}} =$ 10 m/s leeward tower	$U_{\text{wind}} =$ 10 m/s windward tower	$U_{\text{wind}} =$ 10 m/s leeward tower
Total heat exchange rate Q (MW)	248.964	182.956	215.346	186.784	218.272
Relative change ($Q_{\text{wind}} \neq 0 -$ $Q_{\text{wind}} =$ 0)/$Q_{\text{wind}} = 0$ $\times 100\%$		−26.5%	−13.5%	−25.0%	−12.3%

Table 10.4 Comparison of computed results with designed and field-tested values for cooling
towers under no-wind conditions

	Water inlet temperature in heat exchanger T_{wl} (°C)	Environmental air temperature T_{a} (°C)	Water mass flow rate in heat exchanger G (ton/h)	Water outlet temperature in heat exchanger T_{w2} (°C)	Total heat exchange rate Q (MW)
Simulation	43.82	15.46	22,760	32.77	291.89
Design	43.82	15.46	22,760		275.63
Field-test	43.82	15.46	22,760	31.85	316.85

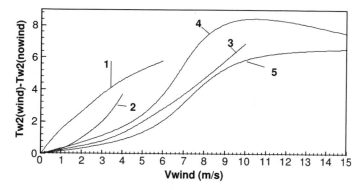

Fig. 10.9 Influence of wind speed on water outlet temperature of heat exchangers of cooling towers: (1) Cooling towers in Russia (Q = 265 MW); (2) Cooling towers in Germany (Q = 188 MW); (3) Cooling tower in Hungary (Q = 331 MW); (4) Predicted windward tower by this study (Q = 200 MW); (5) Predicted leeward tower by this study (Q = 200 MW) [comparison date from the book edited by Ding (1992)]

is less than 10%. Figure 10.9 shows the influence curve of cross-wind speed versus water outlet temperature of heat exchangers. The predicted trends by this study fairly match those measured from actual cooling towers in the world. The predicted internal upward air velocity profiles also show good agreement with the model experiment. The upward airflow speeds at the windward portion of the towers encounter significant reduction and the velocity peak areas are pushed back to the leeward side of the towers. These validations verify that the simulation can provide reasonable results and the results can be used to develop methods for improving the performance of cooling towers.

(3) *Wind influence analysis*

The cooling performance of cooling towers without wind is quite uniform around the towers as evidenced by the uniform water outlet temperatures in Fig. 10.10. The existence of cross-wind of 10 m/s causes 26.5 and 13.5% reduction of the total heat exchange rate for the windward and leeward tower, respectively. This is mainly attributed to the airflow around the cooling towers, destroying the radial flow of surrounding cold air into the towers and thus reducing the heat transfer efficiency of cooling towers. As a result, the water outlet temperatures of the heat exchangers at both lateral sides of the towers (about 0–20°) increase significantly, as shown in Fig. 10.10. Figures 10.11 and 10.12, respectively, display the velocity vector and temperature distributions in the middle section of the cooling towers when U_{wind} = 10 m/s. Figure 10.13 shows the velocity vector distribution in the middle section of the heat exchanger, and Fig. 10.14 illustrates the predicted flow streamlines in and around the towers. All these explicitly indicate that the airflow around the lateral sides of the cooling towers blocks the cold air entering the towers and therefore affects the cooling efficiency of the towers.

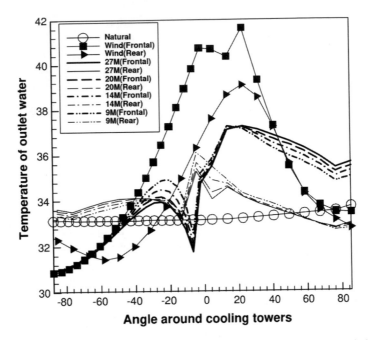

Fig. 10.10 Cooling tower efficiency in terms of water outlet temperature with no wind (natural) and $U_{wind} = 10$ m/s and with different sizes of wind-break walls (X unit: degree; Y unit: °C; −80: windward; 0: lateral; +80: leeward)

Fig. 10.11 Predicted velocity vector distribution in the middle section of the cooling towers when $U_{wind} = 10$ m/s

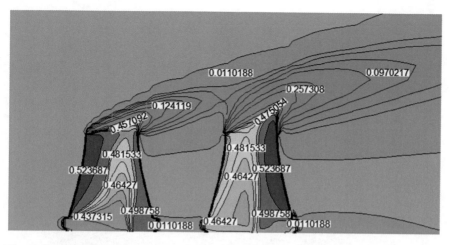

Fig. 10.12 Predicted normalized temperature distribution in the middle section of the cooling towers when $U_{wind} = 10$ m/s

(4) *Improvement performance analysis*

To recover the cooling capacity, wind-break walls were introduced at both sides of the cooling towers, perpendicular to the cross-wind direction. This arrangement will not only hinder the strong cross-flow over the towers but also induce the fresh airflow into the towers through the heat exchangers. The wind-break walls studied were 16 m high (to cover most of the heat exchanger height), 6 m thick (for structure safety concern), and 3.5 m away from the towers (to avoid significant airflow separations at the back of the walls and facilitate maintenance work of the towers). The study compares the performance of the walls with four different widths (9, 14, 20, and 27 m).

The results show that all the wind-break walls can improve the cooling efficiency of the towers under wind conditions. The water outlet temperatures at both sides of cooling towers are reduced, as evidenced in Fig. 10.10. Figure 10.11 presents the airflow patterns at the height of 8.75 m with and without wind-break walls, exhibiting the forced airflows into the towers at the lateral locations by using wind-break walls. The improving effectiveness is increased with the increase of wind-break wall width. But it is not a linear relationship. Figure 10.15 reveals the relationship between cooling tower efficiency recovery rate and the width of wind-break walls. Note that a wider wall does not always improve the tower performance. In fact, it may even make the situation worse because a large separate vortex at the back of a wide wall may block the inflow of air to the towers. The study indicates that the width of 20 m is a good choice for the practical purpose. Various practical forms such as tree walls can be implemented for both blocking cross-wind and cooling surrounding air temperature.

(a)

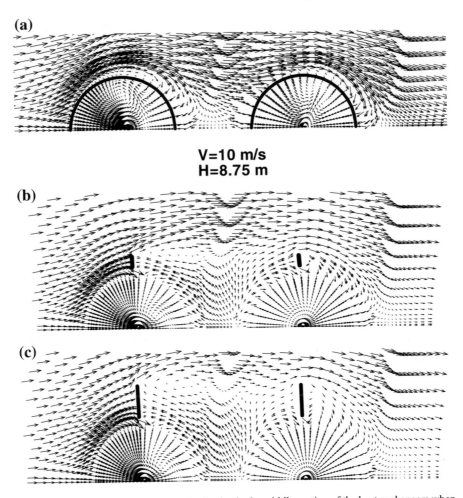

V=10 m/s
H=8.75 m

(b)

(c)

Fig. 10.13 Predicted velocity vector distribution in the middle section of the heat exchangers when $U_{wind} = 10$ m/s: **a** No wind-break walls, **b** 9-m-wide wind-break walls, **c** 27-m-wide wind-break walls

Assignment-10: Overall Review Questions

1. Please list three industrial examples that CFD can and cannot simulate, respectively.
2. Please list three advantages of CFD over experimental fluid dynamics.
3. Please indicate the conventional/practical criteria to define impressible flow.
4. How many equations need be solved in order to obtain room airflow velocity, temperature and humidity distributions?
5. What is Einstein notation (or called summation convention)? Give an example.

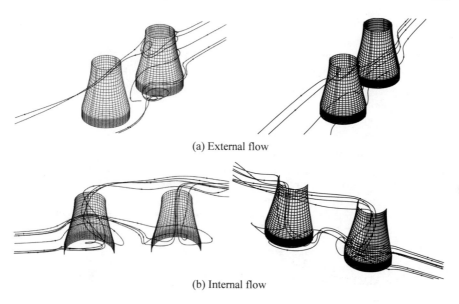

(a) External flow

(b) Internal flow

Fig. 10.14 Predicted flow streamlines in and around the towers when $U_{wind} = 10$ m/s

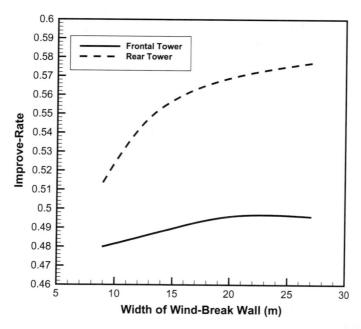

Fig. 10.15 Recovery of cooling tower efficiency at $U_{wind} = 10$ m/s with different sizes of wind-break walls in terms of total heat exchange rate improvement ($Q_{wind/wind-wall} - Q_{wind/no-wind-wall})/(Q_{no-wind} - Q_{wind/no-wind-wall})$

6. How many CFD approaches are available and what are they and which one is usually fastest?
7. What is the Boussinesq Approximation?
8. What is the general expression of a scalar transport equation?
9. Which term do turbulence models deal with in the Reynolds-averaged momentum equations?
10. What is eddy viscosity model?
11. What are the pros and cons of the standard k-e two-equation model?
12. What are low-Reynolds-number k-e models used for?
13. What is symmetric boundary condition?
14. How many velocity boundary conditions are required at a 3-D boundary where the pressure condition is given?
15. How many numerical discretization methods/approaches are available? What are they?
16. What is the upwind differencing scheme?
17. What is TDMA and what for?
18. What is SIMPLE algorithm and what for?
19. What is false-time step?
20. Which one is better: implicit versus explicit algorithm? Why?
21. What are, respectively, staggered grid, structure grid, and adaptive grid?
22. How to judge the grid quality of a structure grid?
23. Please numerically solve the following equation and compare the results with the analytical solution in graph.

$$\frac{dT}{dX} + T = 0,\ 0 \le X \le 1;\ T(0) = 1$$

a. Please test both the upwind and central schemes and comment on how the scheme affects the solutions.
b. Please test different grid resolutions and comment on how the grid affects the solutions.

References

ASHRAE (2013) Guideline 33-2013: guideline for documenting indoor airflow and contaminant transport modeling
ASHRAE (2017) Standard 120-2017: method of testing to determine flow resistance of HVAC ducts and fittings
Ding E (1992) Air cooling techniques in power plants. Water and Electric Power Press, Beijing
Launder BE, Spalding DB (1974) The numerical computation of turbulent flows. Comput Methods Appl Mech Eng 3:269–289
Sleiti A, Zhai Z, Idem S (2013) Computational fluid dynamics to predict duct fitting losses: challenges and opportunities. HVAC&R Res 19(1):2–9

Zhai Z (1999) Study of the flow around dry-cooling towers. Ph.D. Dissertation, Tsinghua University, Beijing, China

Zhai Z, Fu S (2002) Modeling the airflow around cooling towers with multi-block CFD. In: The 4th international ASME/JSME/KSME symposium, Canada

Zhai Z, Fu S (2006) Improving cooling efficiency of dry-cooling towers under cross-wind conditions by using wind-break methods. Appl Therm Eng 26(10):1008–1017